AF541310

ELEMENTARY PHYSICS

ELEMENTARY PHYSICS

– Author –

Pankaj Agnihotri

Virat Singh

Bio-Green Books

New Delhi - 110 002

ISBN: 9788119283200 (Hardback)

Published by : **BIO-GREEN BOOKS**
4736/23, Ansari Road, Darya Ganj,
New Delhi – 110 002
Phone: 011-23245578, 35004336
E-mail: biogreenbooks@gmail.com

Printed at **:** **Replika Press Pvt. Ltd.**

PREFACE

Elementary physics refers to the study of the fundamental concepts and principles of physics, which form the basis of our understanding of the natural world. It covers a wide range of topics, including mechanics, energy, waves, thermodynamics, electricity, and optics. The goal of elementary physics is to provide students with a solid foundation in the subject, which they can build upon as they progress to more advanced topics in physics. It is typically taught at the high school and undergraduate levels and is essential for anyone who wants to pursue a career in physics or a related field. Elementary physics provides a framework for understanding the physical world around us. It helps us explain why objects move, how energy is transferred, and why certain materials behave the way they do.

Many of the technological advancements that we take for granted today, such as smartphones, computers, and the internet, are based on the principles of physics. A solid understanding of elementary physics is essential for the development of new technologies and innovations. Elementary physics is a prerequisite for many careers in science and engineering. It is the foundation for further study in physics and related fields such as engineering, computer science, and astronomy. Elementary physics is also important for understanding the impact of human activities on the environment. It helps us understand how energy is transferred, how heat is distributed, and how materials interact with each other and with the environment. It teaches students how to think logically and systematically when solving problems. This skill is useful not only in science and engineering but also in many other areas of life. An understanding of elementary physics is essential for promoting scientific literacy and understanding scientific concepts in the media and everyday life. Elementary physics is a fundamental subject that is important for understanding the physical world, developing new technologies, pursuing careers in science and engineering, understanding the environment, promoting scientific literacy, and developing problem-solving skills.

The book covers a wide range of topics in physics, including mechanics, simple harmonic motion, work, energy, power, linear momentum, heat, crystal, sound, electric current, and light. The book provides clear and concise explanations of complex physics concepts, making it easy for readers to understand even the most challenging topics. This book is well-organized, with each chapter covering a specific topic and building upon the knowledge gained in the previous chapters. The book includes a variety of exercises and problems at the end of each chapter, allowing readers to test their understanding and apply what they have learned. The book is an excellent resource for students of physics and educators who are looking for a comprehensive yet easy-to-understand textbook.

We are grateful to all those persons as well as various books, manuals, periodicals, magazines, journals etc. that helped in the preparation of this book. In spite of the best efforts, it is possible that some errors may have occurred into the compilation and editing of the book. Further queries, constructive suggestions and criticisms for the improvement of the book are always welcomed and shall be thankfully acknowledged.

Pankaj Agnihotri

Virat Singh

CONTENTS

1

INTRODUCTION

Physics is the natural science that studies matter,[a] its fundamental constituents, its motion and behavior through space and time, and the related entities of energy and force. Physics is one of the most fundamental scientific disciplines, with its main goal being to understand how the universe behaves.[b] A scientist who specializes in the field of physics is called a physicist.

Physics is one of the oldest academic disciplines and, through its inclusion of astronomy, perhaps the oldest. Over much of the past two millennia, physics, chemistry, biology, and certain branches of mathematics were a part of natural philosophy, but during the Scientific Revolution in the 17th century these natural sciences emerged as unique research endeavors in their own right.[c] Physics intersects with many interdisciplinary areas of research, such as biophysics and quantum chemistry, and the boundaries of physics are not rigidly defined. New ideas in physics often explain the fundamental mechanisms studied by other sciences and suggest new avenues of research in these and other academic disciplines such as mathematics and philosophy.

Advances in physics often enable advances in new technologies. For example, advances in the understanding of electromagnetism, solid-state physics, and nuclear physics led directly to the development of new products that have dramatically transformed modern-day society, such as television, computers, domestic appliances, and nuclear weapons; advances in thermodynamics led to the development of industrialization; and advances in mechanics inspired the development of calculus.

Physics, science that deals with the structure of matter and the interactions between the fundamental constituents of the observable universe. In the broadest sense, physics (from the Greek physikos) is concerned with all aspects of nature on both the macroscopic and submicroscopic levels. Its scope of study encompasses not only the behaviour of objects under the action of given forces but also the nature and origin of gravitational, electromagnetic, and nuclear force fields. Its ultimate objective is the formulation of a few comprehensive principles that bring together and explain all such disparate phenomena.

Physics is the basic physical science. Until rather recent times physics and natural philosophy were used interchangeably for the science whose aim is the discovery and formulation of the fundamental laws of nature. As the modern sciences developed and became increasingly specialized, physics came to denote that part of physical science not included in astronomy, chemistry, geology, and engineering. Physics plays an important role in all the natural sciences, however, and all such fields have branches in which physical laws and measurements receive special emphasis, bearing such names as astrophysics, geophysics, biophysics, and even psychophysics. Physics can, at base, be defined as the science of matter, motion, and energy. Its laws are typically expressed with economy and precision in the language of mathematics.

How well do you know astronomy? How about quantum mechanics? This quiz will take you through 36 of the hardest questions from Britannica's most popular quizzes about the sciences. Only the best quizmasters will finish it.

Both experiment, the observation of phenomena under conditions that are controlled as precisely as possible, and theory, the formulation of a unified conceptual framework, play essential and complementary roles in the advancement of physics. Physical experiments result in measurements, which are compared with the outcome predicted by theory. A theory that reliably predicts the results of experiments to which it is applicable is said to embody a law of physics. However, a law is always subject to modification, replacement, or restriction to a more limited domain, if a later experiment makes it necessary.

The ultimate aim of physics is to find a unified set of laws governing matter, motion, and energy at small (microscopic) subatomic distances, at the human (macroscopic) scale of everyday life, and out to the largest distances (e.g., those on the extragalactic scale). This ambitious goal has been realized to a notable extent. Although a completely unified theory of physical phenomena has not yet been achieved (and possibly never will be), a remarkably small set of fundamental physical laws appears able to account for all known phenomena. The body of physics developed up to about the turn of the 20th century, known as classical physics, can largely account for the motions of macroscopic objects that move slowly with respect to the speed of light and for such phenomena as heat, sound, electricity, magnetism, and light. The modern developments of relativity and quantum mechanics modify these laws insofar as they apply to higher speeds, very massive objects, and to the tiny elementary constituents of matter, such as electrons, protons, and neutrons.

HISTORY OF PHYSICS

Physics is a branch of science whose primary objects of study are matter and energy. Discoveries of physics find applications throughout the natural sciences and in technology. Physics today may be divided loosely into classical physics and modern physics.

Ancient history

Elements of what became physics were drawn primarily from the fields of astronomy, optics, and mechanics, which were methodologically united through the study of geometry. These mathematical disciplines began in antiquity with the Babylonians and with Hellenistic writers such as Archimedes and Ptolemy. Ancient philosophy, meanwhile, included what was called "Physics".

Greek concept

The move towards a rational understanding of nature began at least since the Archaic period in Greece (650–480 BCE) with the Pre-Socratic philosophers. The philosopher Thales of Miletus (7th and 6th centuries BCE), dubbed "the Father of Science" for refusing to accept various supernatural, religious or mythological explanations for natural phenomena, proclaimed that every event had a natural cause. Thales also made advancements in 580 BCE by suggesting that water is the basic element, experimenting with the attraction between magnets and rubbed amber and formulating the first recorded cosmologies. Anaximander, famous for his proto-evolutionary theory, disputed Thales' ideas and proposed that rather than water, a substance called apeiron was the building block of all matter. Around 500 BCE, Heraclitus proposed that the only basic law governing the Universe was the principle of change and that nothing remains in the same state indefinitely. This observation made him one of the first scholars in ancient physics to address the role of time in the universe, a key and sometimes contentious concept in modern and present-day physics.

Aristotle

During the classical period in Greece (6th, 5th and 4th centuries BCE) and in Hellenistic times, natural philosophy slowly developed into an exciting and contentious field of study. Aristotle (384 – 322 BCE), a student of Plato, promoted the concept that observation of physical phenomena could ultimately lead to the discovery of the natural laws governing them. Aristotle's writings cover physics, metaphysics, poetry, theater, music, logic, rhetoric, linguistics, politics, government, ethics, biology and zoology. He wrote the first work which refers to that line of study as "Physics" – in the 4th century BCE, Aristotle founded the system known as Aristotelian physics. He attempted to explain ideas such as motion (and gravity) with the theory of four elements. Aristotle believed that all matter was made up of aether, or some combination of four elements: earth, water, air, and fire. According to Aristotle, these four terrestrial elements are capable of inter-transformation and move toward their natural place, so a stone falls downward toward the center of the cosmos, but flames rise upward toward the circumference. Eventually, Aristotelian physics became enormously popular for many centuries in Europe, informing the scientific and scholastic developments of the Middle Ages. It remained the mainstream scientific paradigm in Europe until the time of Galileo Galilei and Isaac Newton.

Early in Classical Greece, knowledge that the Earth is spherical ("round") was common. Around 240 BCE, as the result of a seminal experiment, Eratosthenes (276–194 BCE) accurately estimated its circumference. In contrast to Aristotle's geocentric views, Aristarchus of Samos (Greek: ???sta????; c. 310 – c. 230 BCE) presented an explicit argument for a heliocentric model of the Solar System, i.e. for placing the Sun, not the Earth, at its centre. Seleucus of Seleucia, a follower of Aristarchus' heliocentric theory, stated that the Earth rotated around its own axis, which, in turn, revolved around the Sun. Though the arguments he used were lost, Plutarch stated that Seleucus was the first to prove the heliocentric system through reasoning.

The ancient Greek mathematician Archimedes, famous for his ideas regarding fluid mechanics and buoyancy.

In the 3rd century BCE, the Greek mathematician Archimedes of Syracuse – generally considered to be the greatest mathematician of antiquity and one of the greatest of all time – laid the foundations of hydrostatics, statics and calculated the underlying mathematics of the lever. A leading scientist of classical antiquity, Archimedes also developed elaborate systems of pulleys to move large objects with a minimum of effort. The Archimedes' screw underpins modern hydroengineering, and his machines of war helped to hold back the armies of Rome in the First Punic War. Archimedes even tore apart the arguments of Aristotle and his metaphysics, pointing out that it was impossible to separate mathematics and nature and proved it by converting mathematical theories into practical inventions. Furthermore, in his work On Floating Bodies, around 250 BCE, Archimedes developed the law of buoyancy, also known as Archimedes' principle. In mathematics, Archimedes used the method of exhaustion to calculate the area under the arc of a parabola with the summation of an infinite series, and gave a remarkably accurate approximation of pi. He also defined the spiral bearing his name, formulae for the volumes of surfaces of revolution and an ingenious system for expressing very large numbers. He also developed the principles of equilibrium states and centers of gravity, ideas that would influence the well known scholars, Galileo, and Newton.

Hipparchus (190–120 BCE), focusing on astronomy and mathematics, used sophisticated geometrical techniques to map the motion of the stars and planets, even predicting the times that Solar eclipses would happen. He added calculations of the distance of the Sun and Moon from the Earth, based upon his improvements to the observational instruments used at that time. Another of the most famous of the early physicists was Ptolemy (90–168 CE), one of the leading minds during the time of the Roman Empire. Ptolemy was the author of several scientific treatises, at least three of which were of continuing importance to later Islamic and European science. The first is the astronomical treatise now known as the Almagest. The second is the Geography, which is a thorough discussion of the geographic knowledge of the Greco-Roman world.

Much of the accumulated knowledge of the ancient world was lost. Even of the works of the better known thinkers, few fragments survived. Although he wrote at least fourteen books, almost nothing of Hipparchus' direct work survived. Of the 150 reputed Aristotelian works, only 30 exist, and some of those are "little more than lecture notes".

India and China

In Indian philosophy, Maharishi Kanada was the first to systematically develop a theory of atomism around 200 BCE though some authors have allotted him an earlier era in the 6th century BCE. It was further elaborated by the Buddhist atomists Dharmakirti and Dignaga during the 1st millennium CE. Pakudha Kaccayana, a 6th-century BCE Indian philosopher and contemporary of Gautama Buddha, had also propounded ideas about the atomic constitution of the material world. These philosophers believed that other elements (except ether) were physically palpable and hence comprised minuscule particles of matter. The last minuscule particle of matter that could not be subdivided further was termed Parmanu. These philosophers considered the atom to be indestructible and hence eternal. The Buddhists thought atoms to be minute objects unable to be seen to the naked eye that come into being and vanish in an instant. The Vaisheshika school of philosophers believed that an atom was a mere point in space. It was also first to depict relations between motion and force applied. Indian theories about the atom are greatly abstract and enmeshed in philosophy as they were based on logic and not on personal experience or experimentation. In Indian astronomy, Aryabhata's Aryabhatiya (499 CE) proposed the Earth's rotation, while Nilakantha Somayaji (1444–1544) of the Kerala school of astronomy and mathematics proposed a semi-heliocentric model resembling the Tychonic system.

The study of magnetism in Ancient China dates back to the 4th century BCE. (in the Book of the Devil Valley Master), A main contributor to this field was Shen Kuo (1031–1095), a polymath and statesman who was the first to describe the magnetic-needle compass used for navigation, as well as establishing the concept of true north. In optics, Shen Kuo independently developed a camera obscura.

Islamic world

In the 7th to 15th centuries, scientific progress occurred in the Muslim world. Many classic works in Indian, Assyrian, Sassanian (Persian) and Greek, including the works of Aristotle, were translated into Arabic. Important contributions were made by Ibn al-Haytham (965–1040), an Arab scientist, considered to be a founder of modern optics. Ptolemy and Aristotle theorised that light either shone from the eye to illuminate objects or that "forms" emanated from objects themselves, whereas al-Haytham (known by the Latin name "Alhazen") suggested that light travels to the eye in rays from different points on an object. The works of Ibn al-Haytham and Abu Rayhan Biruni (973–1050), a Persian scientist, eventually passed on to Western Europe where they were studied by scholars such as Roger Bacon and Witelo.

Ibn al-Haytham and Biruni were early proponents of the scientific method. Ibn al-Haytham is considered to be by some the "father of the modern scientific method" due to his emphasis on experimental data and reproducibility of its results. The earliest methodical approach to experiments in the modern sense is visible in the works of Ibn al-Haytham, who introduced an inductive-experimental method for achieving results. Biruni introduced early scientific methods for several different fields of inquiry during the 1020s and 1030s, including an early experimental method for mechanics.[note 2] Biruni's methodology resembled the modern scientific method, particularly in his emphasis on repeated experimentation.

Ibn Sina (980–1037), known as "Avicenna", was a polymath from Bukhara (in present-day Uzbekistan) responsible for important contributions to physics, optics, philosophy and medicine. He published his theory of motion in Book of Healing

(1020), where he argued that an impetus is imparted to a projectile by the thrower, and believed that it was a temporary virtue that would decline even in a vacuum. He viewed it as persistent, requiring external forces such as air resistance to dissipate it. Ibn Sina made a distinction between 'force' and 'inclination' (called "mayl"), and argued that an object gained mayl when the object is in opposition to its natural motion. He concluded that continuation of motion is attributed to the inclination that is transferred to the object, and that object will be in motion until the mayl is spent. He also claimed that projectile in a vacuum would not stop unless it is acted upon. This conception of motion is consistent with Newton's first law of motion, inertia, which states that an object in motion will stay in motion unless it is acted on by an external force. This idea which dissented from the Aristotelian view was later described as "impetus" by John Buridan, who was influenced by Ibn Sina's Book of Healing.

Hibat Allah Abu'l-Barakat al-Baghdaadi (c. 1080-1165) adopted and modified Ibn Sina's theory on projectile motion. In his Kitab al-Mu'tabar, Abu'l-Barakat stated that the mover imparts a violent inclination (mayl qasri) on the moved and that this diminishes as the moving object distances itself from the mover. He also proposed an explanation of the acceleration of falling bodies by the accumulation of successive increments of power with successive increments of velocity. According to Shlomo Pines, al-Baghdaadi's theory of motion was "the oldest negation of Aristotle's fundamental dynamic law [namely, that a constant force produces a uniform motion], [and is thus an] anticipation in a vague fashion of the fundamental law of classical mechanics [namely, that a force applied continuously produces acceleration]." Jean Buridan and Albert of Saxony later referred to Abu'l-Barakat in explaining that the acceleration of a falling body is a result of its increasing impetus.

Ibn Bajjah (c. 1085–1138), known as "Avempace" in Europe, proposed that for every force there is always a reaction force. Ibn Bajjah was a critic of Ptolemy and he worked on creating a new theory of velocity to replace the one theorized by Aristotle. Two future philosophers supported the theories Avempace created, known as Avempacean dynamics. These philosophers were Thomas Aquinas, a Catholic priest, and John Duns Scotus. Galileo went on to adopt Avempace's formula "that the velocity of a given object is the difference of the motive power of that object and the resistance of the medium of motion".

Nasir al-Din al-Tusi (1201–1274), a Persian astronomer and mathematician who died in Baghdad introduced the Tusi couple. Copernicus later drew heavily on the work of al-Din al-Tusi and his students, but without acknowledgment.

Medieval Europe

Awareness of ancient works re-entered the West through translations from Arabic to Latin. Their re-introduction, combined with Judeo-Islamic theological commentaries, had a great influence on Medieval philosophers such as Thomas Aquinas. Scholastic European scholars, who sought to reconcile the philosophy of the ancient classical philosophers with Christian theology, proclaimed Aristotle the greatest thinker of the ancient world. In cases where they didn't directly contradict the Bible, Aristotelian physics became the foundation for the physical explanations of the European Churches. Quantification became a core element of medieval physics.

Based on Aristotelian physics, Scholastic physics described things as moving according to their essential nature. Celestial objects were described as moving in circles, because perfect circular motion was considered an innate property of objects that existed in the uncorrupted realm of the celestial spheres. The theory of impetus, the ancestor to the concepts of inertia and momentum, was developed along similar lines by medieval philosophers such as John Philoponus and Jean Buridan. Motions below the lunar sphere were seen as imperfect, and thus could not be expected to exhibit consistent motion. More idealized motion in the "sublunary" realm could only be achieved through artifice, and prior to the 17th century, many did not view artificial experiments as a valid means of learning about the natural world. Physical explanations in the sublunary realm revolved around tendencies. Stones contained the element earth, and earthly objects tended to move in a straight line toward the centre of the earth (and the universe in the Aristotelian geocentric view) unless otherwise prevented from doing so.

Scientific revolution

During the 16th and 17th centuries, a large advancement of scientific progress known as the Scientific revolution took place in Europe. Dissatisfaction with older philosophical approaches had begun earlier and had produced other changes in society, such as the Protestant Reformation, but the revolution in science began when natural philosophers began to mount a sustained attack on the Scholastic philosophical programme and supposed that mathematical descriptive schemes adopted from such fields as mechanics and astronomy could actually yield universally valid characterizations of motion and other concepts.

Nicolaus Copernicus

A breakthrough in astronomy was made by Polish astronomer Nicolaus Copernicus (1473–1543) when, in 1543, he gave strong arguments for the heliocentric model of the Solar System, ostensibly as a means to render tables charting planetary motion more accurate and to simplify their production. In heliocentric models of the Solar system, the Earth orbits the Sun along with other bodies in Earth's galaxy, a contradiction according to the Greek-Egyptian astronomer Ptolemy (2nd century CE; see above), whose system placed the Earth at the center of the Universe and had been accepted for over 1,400 years. The Greek astronomer Aristarchus of Samos (c.310 – c.230 BCE) had suggested that the Earth revolves around the Sun, but Copernicus' reasoning led to lasting general acceptance of this "revolutionary" idea. Copernicus' book presenting the theory (De revolutionibus orbium coelestium, "On the Revolutions of the Celestial Spheres") was published just before his death in 1543 and, as it is now generally considered to mark the beginning of modern astronomy, is also considered to mark the beginning of the Scientific revolution. Copernicus' new perspective, along with the accurate observations made by Tycho Brahe, enabled German astronomer Johannes Kepler (1571–1630) to formulate his laws regarding planetary motion that remain in use today.

Galileo Galilei

The Italian mathematician, astronomer, and physicist Galileo Galilei (1564–1642) was famous for his support for Copernicanism, his astronomical discoveries, empirical experiments and his improvement of the telescope. As a mathematician, Galileo's role in the university culture of his era was subordinated to the three major topics of study: law, medicine, and theology (which was closely allied to philosophy). Galileo, however, felt that the descriptive content of the technical disciplines warranted philosophical interest, particularly because mathematical analysis of astronomical observations – notably, Copernicus' analysis of the relative motions of the Sun, Earth, Moon, and planets – indicated that philosophers' statements about the nature of the universe could be shown to be in error. Galileo also performed mechanical experiments, insisting that motion itself – regardless of whether it was produced "naturally" or "artificially" (i.e. deliberately) – had universally consistent characteristics that could be described mathematically.

Galileo's early studies at the University of Pisa were in medicine, but he was soon drawn to mathematics and physics. At 19, he discovered (and, subsequently, verified) the isochronal nature of the pendulum when, using his pulse, he timed the oscillations of a swinging lamp in Pisa's cathedral and found that it remained the same for each swing regardless of the swing's amplitude. He soon became known through his invention of a hydrostatic balance and for his treatise on the center of gravity of solid bodies. While teaching at the University of Pisa (1589–92), he initiated his experiments concerning the laws of bodies in motion that brought results so contradictory to the accepted teachings of Aristotle that strong antagonism was aroused. He found that bodies do not fall with velocities proportional to their weights. The famous story in which Galileo is said to have dropped weights from the Leaning Tower of Pisa is apocryphal, but he did find that the path of a projectile is a parabola and is credited with conclusions that anticipated Newton's laws of motion (e.g. the notion of inertia). Among these is what is now called Galilean relativity, the first precisely formulated statement about properties of space and time outside three-dimensional geometry.

A composite montage comparing Jupiter (lefthand side) and its four Galilean moons (top to bottom: Io, Europa, Ganymede, Callisto).

Galileo has been called the "father of modern observational astronomy", the "father of modern physics", the "father of science", and "the father of modern science". According to Stephen Hawking, "Galileo, perhaps more than any other single person, was responsible for the birth of modern science." As religious orthodoxy decreed a geocentric or Tychonic understanding of the Solar system, Galileo's support for heliocentrism provoked controversy and he was tried by the Inquisition. Found "vehemently suspect of heresy", he was forced to recant and spent the rest of his life under house arrest.

The contributions that Galileo made to observational astronomy include the telescopic confirmation of the phases of Venus; his discovery, in 1609, of Jupiter's four largest moons (subsequently given the collective name of the "Galilean moons"); and the observation and analysis of sunspots. Galileo also pursued applied science and technology, inventing, among other instruments, a military compass. His discovery of the Jovian moons was published in 1610 and enabled him to obtain the position of mathematician and philosopher to the Medici court. As such, he was expected to engage in debates with philosophers in the Aristotelian tradition and received a large audience for his own publications such as the Discourses and Mathematical Demonstrations Concerning Two New Sciences (published abroad following his arrest for the publication of Dialogue Concerning the Two Chief World Systems) and The Assayer. Galileo's interest in experimenting with and formulating mathematical descriptions of motion established experimentation as an integral part of natural philosophy. This tradition, combining with the non-mathematical emphasis on the collection of "experimental histories" by philosophical reformists such as William Gilbert and Francis Bacon, drew a significant following in the years leading up to and following Galileo's death, including Evangelista Torricelli and the participants in the Accademia del Cimento in Italy; Marin Mersenne and Blaise

Pascal in France; Christiaan Huygens in the Netherlands; and Robert Hooke and Robert Boyle in England.

RENÉ DESCARTES

The French philosopher René Descartes (1596–1650) was well-connected to, and influential within, the experimental philosophy networks of the day. Descartes had a more ambitious agenda, however, which was geared toward replacing the Scholastic philosophical tradition altogether. Questioning the reality interpreted through the senses, Descartes sought to re-establish philosophical explanatory schemes by reducing all perceived phenomena to being attributable to the motion of an invisible sea of "corpuscles".

(Notably, he reserved human thought and God from his scheme, holding these to be separate from the physical universe). In proposing this philosophical framework, Descartes supposed that different kinds of motion, such as that of planets versus that of terrestrial objects, were not fundamentally different, but were merely different manifestations of an endless chain of corpuscular motions obeying universal principles. Particularly influential were his explanations for circular astronomical motions in terms of the vortex motion of corpuscles in space (Descartes argued, in accord with the beliefs, if not the methods, of the Scholastics, that a vacuum could not exist), and his explanation of gravity in terms of corpuscles pushing objects downward.

Descartes, like Galileo, was convinced of the importance of mathematical explanation, and he and his followers were key figures in the development of mathematics and geometry in the 17th century. Cartesian mathematical descriptions of motion held that all mathematical formulations had to be justifiable in terms of direct physical action, a position held by Huygens and the German philosopher Gottfried Leibniz, who, while following in the Cartesian tradition, developed his own philosophical alternative to Scholasticism, which he outlined in his 1714 work, The Monadology. Descartes has been dubbed the 'Father of Modern Philosophy', and much subsequent Western philosophy is a response to his writings, which are studied closely to this day. In particular, his Meditations on First Philosophy continues to be a standard text at most university philosophy departments. Descartes' influence in mathematics is equally apparent; the Cartesian coordinate system — allowing algebraic equations to be expressed as geometric shapes in a two-dimensional coordinate system — was named after him. He is credited as the father of analytical geometry, the bridge between algebra and geometry, important to the discovery of calculus and analysis.

Christiaan Huygens

The Dutch physicist, mathematician, astronomer and inventor Christiaan Huygens (1629–1695) was the leading scientist in Europe between Galileo and Newton. Huygens came from a family of nobility that had an important position in the Dutch society of the 17th century; a time in which the Dutch Republic flourished economically and culturally. This period - roughly between 1588 and 1702 - of the history of the Netherlands is also referred to as the Dutch Golden Age, an era during the Scientific Revolution when Dutch science was among the most acclaimed in Europe. At this time, intellectuals and scientists like René Descartes, Baruch Spinoza, Pierre Bayle, Antonie van Leeuwenhoek, John Locke and Hugo Grotius resided in the Netherlands. It was in this intellectual environment where Christiaan Huygens grew up. Christiaan's father, Constantijn Huygens, was, apart from an important poet, the secretary and diplomat for the Princes of Orange. He knew many scientists of his time because of his contacts and intellectual interests, including René Descartes and Marin Mersenne, and it was because of these contacts that Christiaan Huygens became aware of their work. Especially Descartes, whose mechanistic philosophy was going to have a huge influence on Huygens' own work. Descartes was later impressed by the skills Christiaan Huygens showed in geometry, as was Mersenne, who christened him "the new Archimedes".

A child prodigy, Huygens began his correspondence with Marin Mersenne when he was 17 years old. Huygens became interested in games of chance when he encountered the work of Fermat, Blaise Pascal and Girard Desargues. It was Blaise Pascal who encourages him to write Van Rekeningh in Spelen van Gluck, which Frans van Schooten translated and published as De Ratiociniis in Ludo Aleae in 1657. The book is the earliest known scientific treatment of the subject, and at the time the most coherent presentation of a mathematical approach to games of chance. Two years later Huygens derived geometrically the now standard formulae in classical mechanics for the centripetal- and centrifugal force in his work De vi Centrifuga (1659). Around the same time Huygens' research in horology resulted in the invention of the pendulum clock; a breakthrough in timekeeping and the most accurate timekeeper for almost 300 years. The theoretical research of the way the pendulum works eventually led to the publication of one of his most important achievements: the Horologium Oscillatorium. This work was published in 1673 and became one of the three most important 17th century works on mechanics (the other two being Galileo's Discourses and Mathematical Demonstrations Relating to Two New Sciences (1638) and Newton's Philosophiæ Naturalis Principia Mathematica (1687)). The Horologium Oscillatorium is the first modern treatise in which a physical problem (the accelerated motion of a falling body) is idealized by a set of parameters then analyzed mathematically and constitutes one of the seminal works of applied mathematics. It is for this reason, Huygens has been called the first theoretical physicist and one of the founders of modern mathematical physics. Huygens' Horologium Oscillatorium had a tremendous influence on the history of physics, especially on the work of Isaac Newton, who greatly admired the work. For instance, the laws Huygens described in the Horologium Oscillatorium are structurally the same as Newton's first two laws of motion.

Five years after the publication of his Horologium Oscillatorium, Huygens described his wave theory of light. Though proposed in 1678, it wasn't published until 1690 in his Traité de la Lumière. His mathematical theory of light was initially rejected in favour of Newton's corpuscular theory of light, until Augustin-Jean Fresnel adopted Huygens' principle to give a complete explanation of the rectilinear propagation and diffraction effects of light in 1821. Today this principle is known as the Huygens–Fresnel principle. As an astronomer, Huygens began grinding lenses with his brother Constantijn jr. to build telescopes for astronomical research. He was the first to identify the rings of Saturn as "a thin, flat ring, nowhere touching, and inclined to the ecliptic," and discovered the first of Saturn's moons, Titan, using a refracting telescope.

Apart from the many important discoveries Huygens made in physics and astronomy, and his inventions of ingenious devices, he was also the first who brought mathematical rigor to the description of physical phenomena. Because of this, and the fact that he developed institutional frameworks for scientific research on the continent, he has been referred to as "the leading actor in 'the making of science in Europe'"

Isaac Newton

The late 17th and early 18th centuries saw the achievements of Cambridge University physicist and mathematician Sir Isaac Newton (1642-1727). Newton, a fellow of the Royal Society of England, combined his own discoveries in mechanics and astronomy to earlier ones to create a single system for describing the workings of the universe. Newton formulated three laws of motion which formulated the relationship between motion and objects and also the law of universal gravitation, the latter of which could be used to explain the behavior not only of falling bodies on the earth but also planets and other celestial bodies. To arrive at his results, Newton invented one form of an entirely new branch of mathematics: calculus (also invented independently by Gottfried Leibniz), which was to become an essential tool in much of the later development in most branches of physics. Newton's findings were set forth in his Philosophiæ Naturalis Principia Mathematica ("Mathematical Principles of Natural Philosophy"), the publication of which in 1687 marked the beginning of the modern period of mechanics and astronomy.

Newton was able to refute the Cartesian mechanical tradition that all motions should be explained with respect to the immediate force exerted by corpuscles. Using his three laws of motion and law of universal gravitation, Newton removed the idea that objects followed paths determined by natural shapes and instead demonstrated that not only regularly observed paths, but all the future motions of any body could be deduced mathematically based on knowledge of their existing motion, their mass, and the forces acting upon them. However, observed celestial motions did not precisely conform to a Newtonian treatment, and Newton, who was also deeply interested in theology, imagined that God intervened to ensure the continued stability of the solar system.

Gottfried Leibniz

Newton's principles proved controversial with Continental philosophers, who found his lack of metaphysical explanation for movement and gravitation philosophically unacceptable. Beginning around 1700, a bitter rift opened between the Continental and British philosophical traditions, which were stoked by heated, ongoing, and viciously personal disputes between the followers of Newton and Leibniz concerning priority over the analytical techniques of calculus, which each had developed independently. Initially, the Cartesian and Leibnizian traditions prevailed on the Continent (leading to the dominance of the Leibnizian calculus notation everywhere except Britain). Newton himself remained privately disturbed at the lack of a philosophical understanding of gravitation while insisting in his writings that none was necessary to infer its reality. As the 18th century progressed, Continental natural philosophers increasingly accepted the Newtonians' willingness to forgo ontological metaphysical explanations for mathematically described motions.

Newton built the first functioning reflecting telescope and developed a theory of color, published in Opticks, based on the observation that a prism decomposes white light into the many colours forming the visible spectrum. While Newton explained light as being composed of tiny particles, a rival theory of light which explained its behavior in terms of waves was presented in 1690 by Christiaan Huygens. However, the belief in the mechanistic philosophy coupled with Newton's reputation meant that the wave theory saw relatively little support until the 19th century. Newton also formulated an empirical law of cooling, studied the speed of sound, investigated power series, demonstrated the generalised binomial theorem and developed a method for approximating the roots of a function. His work on infinite series was inspired by Simon Stevin's decimals. Most importantly, Newton showed that the motions of objects on Earth and of celestial bodies are governed by the same set of natural laws, which were neither capricious nor malevolent. By demonstrating the consistency between Kepler's laws of planetary motion and his own theory of gravitation, Newton also removed the last doubts about heliocentrism. By bringing together all the ideas set forth during the Scientific revolution, Newton effectively established the foundation for modern society in mathematics and science.

Other Achievements

Other branches of physics also received attention during the period of the Scientific revolution. William Gilbert, court physician to Queen Elizabeth I, published an important work on magnetism in 1600, describing how the earth itself behaves like a giant magnet. Robert Boyle (1627–91) studied the behavior of gases enclosed in a chamber and formulated the gas law named for him; he also contributed to physiology and to the founding of modern chemistry. Another important factor in the scientific revolution was the rise of learned societies and academies in various countries. The earliest of these were in Italy and Germany and were short-lived. More influential were the Royal Society of England (1660) and the Academy of Sciences in France (1666). The former was a private institution in London and included such scientists as John Wallis, William Brouncker, Thomas Sydenham, John Mayow, and Christopher Wren (who contributed not only to architecture but also to astronomy and anatomy); the latter, in Paris, was a government institution and included as a foreign member the Dutchman Huygens. In the 18th century, important royal academies were established at Berlin (1700) and at St. Petersburg (1724). The societies and academies provided the principal opportunities for the publication and discussion of scientific results during and after the scientific revolution. In 1690, James Bernoulli showed that the cycloid is the solution to the tautochrone problem; and the following year, in 1691, Johann Bernoulli showed that a chain freely suspended from two points will form a catenary, the curve with the lowest possible center of gravity available to any chain hung between two fixed points. He then showed, in 1696, that the cycloid is the solution to the brachistochrone problem.

Early Thermodynamics

A precursor of the engine was designed by the German scientist Otto von Guericke who, in 1650, designed and built the world's first vacuum pump to create a vacuum as demonstrated in the Magdeburg hemispheres experiment. He was driven to make a vacuum to disprove Aristotle's long-held supposition that 'Nature abhors a vacuum'. Shortly thereafter, Irish physicist and chemist Boyle had learned of Guericke's designs and in 1656, in coordination with English scientist Robert Hooke, built an air pump. Using this pump, Boyle and Hooke noticed the pressure-volume correlation for a gas: PV = k, where P is pressure, V is volume and k is a constant: this relationship is known as Boyle's Law. In that time, air was assumed to be a system of motionless particles, and not interpreted as a system of moving molecules. The concept of thermal motion came two centuries later. Therefore, Boyle's publication in 1660 speaks about a mechanical concept: the air spring. Later, after the invention of the thermometer, the property temperature could be quantified. This tool gave Gay-Lussac the opportunity to derive his law, which led shortly later to the ideal gas law. But, already before the establishment of the ideal gas law, an associate of Boyle's named Denis Papin built in 1679 a bone digester, which is a closed vessel with a tightly fitting lid that confines steam until a high pressure is generated.

Later designs implemented a steam release valve to keep the machine from exploding. By watching the valve rhythmically move up and down, Papin conceived of the idea of a piston and cylinder engine. He did not however follow through with his design. Nevertheless, in 1697, based on Papin's designs, engineer Thomas Savery built the first engine. Although these early engines were crude and inefficient, they attracted the attention of the leading scientists of the time. Hence, prior to 1698 and the invention of the Savery Engine, horses were used to power pulleys, attached to buckets, which lifted water out of flooded salt mines in England. In the years to follow, more variations of steam engines were built, such as the Newcomen Engine, and later the Watt Engine. In time, these early engines would eventually be utilized in place of horses. Thus, each engine began to be associated with a certain amount of "horse power" depending upon how many horses it had replaced. The main problem with these first engines was that they were slow and clumsy, converting less than 2% of the input fuel into useful work. In other words, large quantities of coal (or wood) had to be burned to yield only a small fraction of work output. Hence the need for a new science of engine dynamics was born.

18th-century developments

Alessandro Volta

During the 18th century, the mechanics founded by Newton was developed by several scientists as more mathematicians learned calculus and elaborated upon its initial formulation. The application of mathematical analysis to problems of motion was known as rational mechanics, or mixed mathematics.

Daniel Bernoulli

In 1714, Brook Taylor derived the fundamental frequency of a stretched vibrating string in terms of its tension and mass per unit length by solving a differential equation. The Swiss mathematician Daniel Bernoulli (1700–1782) made important mathematical studies of the behavior of gases, anticipating the kinetic theory of gases developed more than a century later, and has been referred to as the first mathematical physicist. In 1733, Daniel Bernoulli derived the fundamental frequency and harmonics of a hanging chain by solving a differential equation. In 1734, Bernoulli solved the differential equation for the vibrations of an elastic bar clamped at one end. Bernoulli's treatment of fluid dynamics and his examination of fluid flow was introduced in his 1738 work Hydrodynamica.

Rational mechanics dealt primarily with the development of elaborate mathematical treatments of observed motions, using Newtonian principles as a basis, and emphasized improving the tractability of complex calculations and developing of legitimate means of analytical approximation. A representative contemporary textbook was published by Johann Baptiste Horvath. By the end of the century analytical treatments were rigorous enough to verify the stability of the Solar System solely on the basis of Newton's laws without reference to divine intervention—even as deterministic treatments of systems as simple as the three body problem in gravitation remained intractable. In 1705, Edmond Halley predicted the periodicity of Halley's Comet, William Herschel discovered Uranus in 1781, and Henry Cavendish measured the gravitational constant and determined the mass of the Earth in 1798. In 1783, John Michell suggested that some objects might be so massive that not even light could escape from them.

In 1739, Leonhard Euler solved the ordinary differential equation for a forced harmonic oscillator and noticed the resonance phenomenon. In 1742, Colin Maclaurin discovered his uniformly rotating self-gravitating spheroids. In 1742, Benjamin Robins published his New Principles in Gunnery, establishing the science of aerodynamics. British work, carried on by mathematicians such as Taylor and Maclaurin, fell behind Continental developments as the century progressed. Meanwhile, work flourished at scientific academies on the Continent, led by such mathematicians as Bernoulli, Euler, Lagrange, Laplace, and Legendre. In 1743, Jean le Rond d'Alembert published his Traite de Dynamique, in which he introduced the concept of generalized forces for accelerating systems and systems with constraints, and applied the new idea of virtual work to solve dynamical problem, now known as

D'Alembert's principle, as a rival to Newton's second law of motion. In 1747, Pierre Louis Maupertuis applied minimum principles to mechanics. In 1759, Euler solved the partial differential equation for the vibration of a rectangular drum. In 1764, Euler examined the partial differential equation for the vibration of a circular drum and found one of the Bessel function solutions. In 1776, John Smeaton published a paper on experiments relating power, work, momentum and kinetic energy, and supporting the conservation of energy. In 1788, Joseph Louis Lagrange presented Lagrange's equations of motion in Mécanique Analytique, in which the whole of mechanics was organized around the principle of virtual work. In 1789, Antoine Lavoisier states the law of conservation of mass. The rational mechanics developed in the 18th century received a brilliant exposition in both Lagrange's 1788 work and the Celestial Mechanics (1799–1825) of Pierre-Simon Laplace.

Thermodynamics

During the 18th century, thermodynamics was developed through the theories of weightless "imponderable fluids", such as heat ("caloric"), electricity, and phlogiston (which was rapidly overthrown as a concept following Lavoisier's identification of oxygen gas late in the century). Assuming that these concepts were real fluids, their flow could be traced through a mechanical apparatus or chemical reactions. This tradition of experimentation led to the development of new kinds of experimental apparatus, such as the Leyden Jar; and new kinds of measuring instruments, such as the calorimeter, and improved versions of old ones, such as the thermometer. Experiments also produced new concepts, such as the University of Glasgow experimenter Joseph Black's notion of latent heat and Philadelphia intellectual Benjamin Franklin's characterization of electrical fluid as flowing between places of excess and deficit (a concept later reinterpreted in terms of positive and negative charges). Franklin also showed that lightning is electricity in 1752.

The accepted theory of heat in the 18th century viewed it as a kind of fluid, called caloric; although this theory was later shown to be erroneous, a number of scientists adhering to it nevertheless made important discoveries useful in developing the modern theory, including Joseph Black (1728–99) and Henry Cavendish (1731–1810). Opposed to this caloric theory, which had been developed mainly by the chemists, was the less accepted theory dating from Newton's time that heat is due to the motions of the particles of a substance. This mechanical theory gained support in 1798 from the cannon-boring experiments of Count Rumford (Benjamin Thompson), who found a direct relationship between heat and mechanical energy.

While it was recognized early in the 18th century that finding absolute theories of electrostatic and magnetic force akin to Newton's principles of motion would be an important achievement, none were forthcoming. This impossibility only slowly disappeared as experimental practice became more widespread and more refined in the early years of the 19th century in places such as the newly established Royal Institution in London. Meanwhile, the analytical methods of rational mechanics began to be applied to experimental phenomena, most influentially with the French mathematician Joseph Fourier's analytical treatment of the flow of heat, as published in 1822. Joseph Priestley proposed an electrical inverse-square law in 1767, and Charles-Augustin de Coulomb introduced the inverse-square law of electrostatics in 1798.

At the end of the century, the members of the French Academy of Sciences had attained clear dominance in the field. At the same time, the experimental tradition established by Galileo and his followers persisted. The Royal Society and the French Academy of Sciences were major centers for the performance and reporting of experimental work. Experiments in mechanics, optics, magnetism, static electricity, chemistry, and physiology were not clearly distinguished from each other during the 18th century, but significant differences in explanatory schemes and, thus, experiment design were emerging. Chemical experimenters, for instance, defied attempts to enforce a scheme of abstract Newtonian forces onto chemical affiliations, and instead focused on the isolation and classification of chemical substances and reactions.

19th century

In 1813, Peter Ewart supported the idea of the conservation of energy in his paper On the measure of moving force. In 1829, Gaspard Coriolis introduced the terms of work (force times distance) and kinetic energy with the meanings they have today. In 1841, Julius Robert von Mayer, an amateur scientist, wrote a paper on the conservation of energy, although his lack of academic training led to its rejection. In 1847, Hermann von Helmholtz formally stated the law of conservation of energy.

Electromagnetism

In 1800, Alessandro Volta invented the electric battery (known as the voltaic pile) and thus improved the way electric currents could also be studied. A year later, Thomas Young demonstrated the wave nature of light—which received strong experimental support from the work of Augustin-Jean Fresnel—and the principle of interference. In 1820, Hans Christian Ørsted found that a current-carrying conductor gives rise to a magnetic force surrounding it, and within a week after Ørsted's discovery reached France, André-Marie Ampère discovered that two parallel electric currents will exert forces on each other. In 1821, Michael Faraday built an electricity-powered motor, while Georg Ohm stated his law of electrical resistance in 1826, expressing the relationship between voltage, current, and resistance in an electric circuit.

In 1831, Faraday (and independently Joseph Henry) discovered the reverse effect, the production of an electric potential or current through magnetism - known as electromagnetic induction; these two discoveries are the basis of the electric motor and the electric generator, respectively.

LAWS OF THERMODYNAMICS

In the 19th century, the connection between heat and mechanical energy was established quantitatively by Julius Robert von Mayer and James Prescott Joule, who measured the mechanical equivalent of heat in the 1840s. In 1849, Joule published results from his series of experiments (including the paddlewheel experiment) which show that heat is a form of energy, a fact that was accepted in the 1850s. The relation between heat and energy was important for the development of steam engines, and in 1824 the experimental and theoretical work of Sadi Carnot was published. Carnot captured some of the ideas of thermodynamics in his discussion of the efficiency of an idealized engine. Sadi Carnot's work provided a basis for the formulation of the first law of thermodynamics—a restatement of the law of conservation of energy—which was stated around 1850 by William Thomson, later known as Lord Kelvin, and Rudolf Clausius. Lord Kelvin, who had extended the concept of absolute zero from gases to all substances in 1848, drew upon the engineering theory of Lazare Carnot, Sadi Carnot, and Émile Clapeyron–as well as the experimentation of James Prescott Joule on the interchangeability of mechanical, chemical, thermal, and electrical forms of work—to formulate the first law.

Kelvin and Clausius also stated the second law of thermodynamics, which was originally formulated in terms of the fact that heat does not spontaneously flow from a colder body to a hotter. Other formulations followed quickly (for example, the second law was expounded in Thomson and Peter Guthrie Tait's influential work Treatise on Natural Philosophy) and Kelvin in particular understood some of the law's general implications. The second Law was the idea that gases consist of molecules in motion had been discussed in some detail by Daniel Bernoulli in 1738, but had fallen out of favor, and was revived by Clausius in 1857. In 1850, Hippolyte Fizeau and Léon Foucault measured the speed of light in water and find that it is slower than in air, in support of the wave model of light. In 1852, Joule and Thomson demonstrated that a rapidly expanding gas cools, later named the Joule–Thomson effect or Joule–Kelvin effect. Hermann von Helmholtz puts forward the idea of the heat death of the universe in 1854, the same year that Clausius established the importance of dQ/T (Clausius's theorem) (though he did not yet name the quantity).

Statistical Mechanics

In 1859, James Clerk Maxwell discovered the distribution law of molecular velocities. Maxwell showed that electric and magnetic fields are propagated outward from their source at a speed equal to that of light and that light is one of several kinds of electromagnetic radiation, differing only in frequency and wavelength from the others. In 1859, Maxwell worked out the mathematics of the distribution of velocities of the molecules of a gas. The wave theory of light was widely accepted by the time of Maxwell's work on the electromagnetic field, and afterward the study of light and that of electricity and magnetism were closely related. In 1864 James Maxwell published his papers on a dynamical theory of the electromagnetic field, and stated that light is an electromagnetic phenomenon in the 1873 publication of Maxwell's Treatise on Electricity and Magnetism. This work drew upon theoretical work by German theoreticians such as Carl Friedrich Gauss and Wilhelm Weber. The encapsulation of heat in particulate motion, and the addition of electromagnetic forces to Newtonian dynamics established an enormously robust theoretical underpinning to physical observations.

The prediction that light represented a transmission of energy in wave form through a "luminiferous ether", and the seeming confirmation of that prediction with Helmholtz student Heinrich Hertz's 1888 detection of electromagnetic radiation, was a major triumph for physical theory and raised the possibility that even more fundamental theories based on the field could soon be developed. Experimental confirmation of Maxwell's theory was provided by Hertz, who generated and detected electric waves in 1886 and verified their properties, at the same time foreshadowing their application in radio, television, and other devices. In 1887, Heinrich Hertz discovered the photoelectric effect. Research on the electromagnetic waves began soon after, with many scientists and inventors conducting experiments on their properties. In the mid to late 1890s Guglielmo Marconi developed a radio wave based wireless telegraphy system.

The atomic theory of matter had been proposed again in the early 19th century by the chemist John Dalton and became one of the hypotheses of the kinetic-molecular theory of gases developed by Clausius and James Clerk Maxwell to explain the laws of thermodynamics.

Ludwig Boltzmann

The kinetic theory in turn led to a revolutionary approach to science, the statistical mechanics of Ludwig Boltzmann (1844–1906) and Josiah Willard Gibbs (1839–1903), which studies the statistics of microstates of a system and uses statistics to determine the state of a physical system. Interrelating the statistical likelihood of certain states of organization of these particles with the energy of those states, Clausius reinterpreted the dissipation of energy to be the statistical tendency of molecular configurations to pass toward increasingly likely, increasingly disorganized states (coining the term "entropy" to describe the disorganization of a state). The statistical versus absolute interpretations of the second law of thermodynamics set up a dispute that would last for several decades (producing arguments such as "Maxwell's demon"), and that would not be held to be definitively resolved until the behavior of atoms was firmly established in the early 20th century. In 1902, James Jeans found the length scale required for gravitational perturbations to grow in a static nearly homogeneous medium.

Other Developments

In 1822, botanist Robert Brown discovered Brownian motion: pollen grains in water undergoing movement resulting from their bombardment by the fast-moving atoms or molecules in the liquid.

In 1834, Carl Jacobi discovered his uniformly rotating self-gravitating ellipsoids (the Jacobi ellipsoid).

In 1834, John Russell observed a nondecaying solitary water wave (soliton) in the Union Canal near Edinburgh and used a water tank to study the dependence of solitary water wave velocities on wave amplitude and water depth. In 1835, Gaspard Coriolis examined theoretically the mechanical efficiency of waterwheels, and deduced the Coriolis effect. In 1842, Christian Doppler proposed the Doppler effect.

In 1851, Léon Foucault showed the Earth's rotation with a huge pendulum (Foucault pendulum).

There were important advances in continuum mechanics in the first half of the century, namely formulation of laws of elasticity for solids and discovery of Navier–Stokes equations for fluids.

20th century: birth of modern physics

At the end of the 19th century, physics had evolved to the point at which classical mechanics could cope with highly complex problems involving macroscopic situations; thermodynamics and kinetic theory were well established; geometrical and physical optics could be understood in terms of electromagnetic waves; and the conservation laws for energy and momentum (and mass) were widely accepted. So profound were these and other developments that it was generally accepted that all the important laws of physics had been discovered and that, henceforth, research would be concerned with clearing up minor problems and particularly with improvements of method and measurement. However, around 1900 serious doubts arose about the completeness of the classical theories—the triumph of Maxwell's theories, for example, was undermined by inadequacies that had already begun to appear—and their inability to explain certain physical phenomena, such as the energy distribution in blackbody radiation and the photoelectric effect, while some of the theoretical formulations led to paradoxes when pushed to the limit. Prominent physicists such as Hendrik Lorentz, Emil Cohn, Ernst Wiechert and Wilhelm Wien believed that some modification of Maxwell's equations might provide the basis for all physical laws. These shortcomings of classical physics were never to be resolved and new ideas were required. At the beginning of the 20th century a major revolution shook the world of physics, which led to a new era, generally referred to as modern physics.

Radiation experiments

In the 19th century, experimenters began to detect unexpected forms of radiation: Wilhelm Röntgen caused a sensation with his discovery of X-rays in 1895; in 1896 Henri Becquerel discovered that certain kinds of matter emit radiation on their own accord. In 1897, J. J. Thomson discovered the electron, and new radioactive elements found by Marie and Pierre Curie raised questions about the supposedly indestructible atom and the nature of matter. Marie and Pierre coined the term "radioactivity" to describe this property of matter, and isolated the radioactive elements radium and polonium. Ernest Rutherford and Frederick Soddy identified two of Becquerel's forms of radiation with electrons and the element helium. Rutherford identified and named two types of radioactivity and in 1911 interpreted experimental evidence as showing that the atom consists of a dense, positively charged nucleus surrounded by negatively charged electrons. Classical theory, however, predicted that this structure should be unstable. Classical theory had also failed to explain successfully two other experimental results that appeared in the late 19th century. One of these was the demonstration by Albert A. Michelson and Edward W. Morley—known as the Michelson–Morley experiment—which showed there did not seem to be a preferred frame of reference, at rest with respect to the hypothetical luminiferous ether, for describing electromagnetic phenomena.

Studies of radiation and radioactive decay continued to be a preeminent focus for physical and chemical research through the 1930s, when the discovery of nuclear fission by Lise Meitner and Otto Frisch opened the way to the practical exploitation of what came to be called "atomic" energy.

ALBERT EINSTEIN'S THEORY OF RELATIVITY

In 1905, a 26-year-old German physicist named Albert Einstein (then a patent clerk in Bern, Switzerland) showed how measurements of time and space are affected by motion between an observer and what is being observed. Einstein's radical theory of relativity revolutionized science. Although Einstein made many other important contributions to science, the theory of relativity alone represents one of the greatest intellectual achievements of all time. Although the concept of relativity was not introduced by Einstein, his major contribution was the recognition that the speed of light in a vacuum is constant, i.e. the same for all observers, and an absolute physical boundary for motion. This does not impact a person's day-to-day life since most objects travel at speeds much slower than light speed. For objects travelling near light speed, however, the theory of relativity shows that clocks associated with those objects will run more slowly and that the objects shorten in length according to measurements of an observer on Earth. Einstein also derived the famous equation, E = mc2, which expresses the equivalence of mass and energy.

Special relativity

Einstein proposed that gravitation is a result of masses (or their equivalent energies) curving ("bending") the spacetime in which they exist, altering the paths they follow within it.

Einstein argued that the speed of light was a constant in all inertial reference frames and that electromagnetic laws should remain valid independent of reference frame—assertions which rendered the ether "superfluous" to physical theory, and that held that observations of time and length varied relative to how the observer was moving with respect to the object being measured (what came to be called the "special theory of relativity"). It also followed that mass and energy were interchangeable quantities according to the equation E=mc2. In another paper published the same year, Einstein asserted that electromagnetic radiation was transmitted in discrete quantities ("quanta"), according to a constant that the theoretical physicist Max Planck had posited in 1900 to arrive at an accurate theory for the distribution of blackbody radiation—an assumption that explained the strange properties of the photoelectric effect.

The special theory of relativity is a formulation of the relationship between physical observations and the concepts of space and time. The theory arose out of contradictions between electromagnetism and Newtonian mechanics and had great impact on both those areas. The original historical issue was whether it was meaningful to discuss the electromagnetic wave-carrying "ether" and motion relative to it and also whether one could detect such motion, as was unsuccessfully attempted in the Michelson–Morley experiment. Einstein demolished these questions and the ether concept in his special theory of relativity.

However, his basic formulation does not involve detailed electromagnetic theory. It arises out of the question: "What is time?" Newton, in the Principia (1686), had given an unambiguous answer: "Absolute, true, and mathematical time, of itself, and from its own nature, flows equably without relation to anything external, and by another name is called duration." This definition is basic to all classical physics.

Einstein had the genius to question it, and found that it was incomplete. Instead, each "observer" necessarily makes use of his or her own scale of time, and for two observers in relative motion, their time-scales will differ. This induces a related effect on position measurements. Space and time become intertwined concepts, fundamentally dependent on the observer. Each observer presides over his or her own space-time framework or coordinate system. There being no absolute frame of reference, all observers of given events make different but equally valid (and reconcilable) measurements. What remains absolute is stated in Einstein's relativity postulate: "The basic laws of physics are identical for two observers who have a constant relative velocity with respect to each other."

Special relativity had a profound effect on physics: started as a rethinking of the theory of electromagnetism, it found a new symmetry law of nature, now called Poincaré symmetry, that replaced the old Galilean symmetry.

Special relativity exerted another long-lasting effect on dynamics. Although initially it was credited with the "unification of mass and energy", it became evident that relativistic dynamics established a firm distinction between rest mass, which is an invariant (observer independent) property of a particle or system of particles, and the energy and momentum of a system. The latter two are separately conserved in all situations but not invariant with respect to different observers. The term mass in particle physics underwent a semantic change, and since the late 20th century it almost exclusively denotes the rest (or invariant) mass.

General Relativity

By 1916, Einstein was able to generalize this further, to deal with all states of motion including non-uniform acceleration, which became the general theory of relativity. In this theory Einstein also specified a new concept, the curvature of space-time, which described the gravitational effect at every point in space. In fact, the curvature of space-time completely replaced Newton's universal law of gravitation. According to Einstein, gravitational force in the normal sense is a kind of illusion caused by the geometry of space. The presence of a mass causes a curvature of space-time in the vicinity of the mass, and this curvature dictates the space-time path that all freely-moving objects must follow. It was also predicted from this theory that light should be subject to gravity - all of which was verified experimentally. This aspect of relativity explained the phenomena of light bending around the sun, predicted black holes as well as properties of the Cosmic microwave background radiation — a discovery rendering fundamental anomalies in the classic Steady-State hypothesis. For his work on relativity, the photoelectric effect and blackbody radiation, Einstein received the Nobel Prize in 1921.

The gradual acceptance of Einstein's theories of relativity and the quantized nature of light transmission, and of Niels Bohr's model of the atom created as many problems as they solved, leading to a full-scale effort to reestablish physics on new fundamental principles. Expanding relativity to cases of accelerating reference frames (the "general theory of relativity") in the 1910s, Einstein posited an equivalence between the inertial force of acceleration and the force of gravity, leading to the conclusion that space is curved and finite in size, and the prediction of such phenomena as gravitational lensing and the distortion of time in gravitational fields.

Quantum Mechanics

Although relativity resolved the electromagnetic phenomena conflict demonstrated by Michelson and Morley, a second theoretical problem was the explanation of the distribution of electromagnetic radiation emitted by a black body; experiment showed that at shorter wavelengths, toward the ultraviolet end of the spectrum, the energy approached zero, but classical theory predicted it should become infinite. This glaring discrepancy, known as the ultraviolet catastrophe, was solved by the new theory of quantum mechanics. Quantum mechanics is the theory of atoms and subatomic systems. Approximately the first 30 years of the 20th century represent the time of the conception and evolution of the theory. The basic ideas of quantum theory were introduced in 1900 by Max Planck (1858–1947), who was awarded the Nobel Prize for Physics in 1918 for his discovery of the quantified nature of energy. The quantum theory (which previously relied in the "correspondence" at large scales between the quantized world of the atom and the continuities of the "classical" world) was accepted when the Compton Effect established that light carries momentum and can scatter off particles, and when Louis de Broglie asserted that matter can be seen as behaving as a wave in much the same way as electromagnetic waves behave like particles (wave–particle duality).

Werner Heisenberg

In 1905, Einstein used the quantum theory to explain the photoelectric effect, and in 1913 the Danish physicist Niels Bohr used the same constant to explain the stability of Rutherford's atom as well as the frequencies of light emitted by hydrogen gas. The quantized theory of the atom gave way to a full-scale quantum mechanics in the 1920s.

New principles of a "quantum" rather than a "classical" mechanics, formulated in matrix-form by Werner Heisenberg, Max Born, and Pascual Jordan in 1925, were based on the probabilistic relationship between discrete "states" and denied the possibility of causality. Quantum mechanics was extensively developed by Heisenberg, Wolfgang Pauli, Paul Dirac, and Erwin Schrödinger, who established an equivalent theory based on waves in 1926; but Heisenberg's 1927 "uncertainty principle" (indicating the impossibility of precisely and simultaneously measuring position and momentum) and the "Copenhagen interpretation" of quantum mechanics (named after Bohr's home city) continued to deny the possibility of fundamental causality, though opponents such as Einstein would metaphorically assert that "God does not play dice with the universe".

The new quantum mechanics became an indispensable tool in the investigation and explanation of phenomena at the atomic level. Also in the 1920s, the Indian scientist Satyendra Nath Bose's work on photons and quantum mechanics provided the foundation for Bose–Einstein statistics, the theory of the Bose–Einstein condensate.

The spin–statistics theorem established that any particle in quantum mechanics may be either a boson (statistically Bose–Einstein) or a fermion (statistically Fermi–Dirac). It was later found that all fundamental bosons transmit forces, such as the photon that transmits electromagnetism.

Fermions are particles "like electrons and nucleons" and are the usual constituents of matter. Fermi–Dirac statistics later found numerous other uses, from astrophysics (see Degenerate matter) to semiconductor design.

CONTEMPORARY AND PARTICLE PHYSICS

Quantum field theory

A Feynman diagram representing (left to right) the production of a photon (blue sine wave) from the annihilation of an electron and its complementary antiparticle, the positron. The photon becomes a quark–antiquark pair and a gluon (green spiral) is released.

As the philosophically inclined continued to debate the fundamental nature of the universe, quantum theories continued to be produced, beginning with Paul Dirac's formulation of a relativistic quantum theory in 1928. However, attempts to quantize electromagnetic theory entirely were stymied throughout the 1930s by theoretical formulations yielding infinite energies. This situation was not considered adequately resolved until after World War II ended, when Julian Schwinger, Richard Feynman and Sin-Itiro Tomonaga independently posited the technique of renormalization, which allowed for an establishment of a robust quantum electrodynamics (QED).

Meanwhile, new theories of fundamental particles proliferated with the rise of the idea of the quantization of fields through "exchange forces" regulated by an exchange of short-lived "virtual" particles, which were allowed to exist according to the laws governing the uncertainties inherent in the quantum world. Notably, Hideki Yukawa proposed that the positive charges of the nucleus were kept together courtesy of a powerful but short-range force mediated by a particle with a mass between that of the electron and proton. This particle, the "pion", was identified in 1947 as part of what became a slew of particles discovered after World War II. Initially, such particles were found as ionizing radiation left by cosmic rays, but increasingly came to be produced in newer and more powerful particle accelerators.

the theoretical and experimental research of superconductivity, especially the invention of a quantum theory of superconductivity by Vitaly Ginzburg and Lev Landau (1962 Nobel Prize in Physics) and, later, its explanation via Cooper pairs (1972 Nobel Prize in Physics). The Cooper pair was an early example of quasiparticles.

Unified Field Theories

Einstein deemed that all fundamental interactions in nature can be explained in a single theory. Unified field theories were numerous attempts to "merge" several interactions. One of many formulations of such theories (as well as field theories in general) is a gauge theory, a generalization of the idea of symmetry. Eventually the Standard Model (see below) succeeded in unification of strong, weak, and electromagnetic interactions. All attempts to unify gravitation with something else failed.

The Standard Model

When parity was broken in weak interactions by Chien-Shiung Wu in her experiment, a series of discoveries were created thereafter. The interaction of these particles by scattering and decay provided a key to new fundamental quantum theories. Murray Gell-Mann and Yuval Ne'eman brought some order to these new particles by classifying them according to certain qualities, beginning with what Gell-Mann referred to as the "Eightfold Way". While its further development, the quark model, at first seemed inadequate to describe strong nuclear forces, allowing the temporary rise of competing theories such as the S-Matrix, the establishment of quantum chromodynamics in the 1970s finalized a set of fundamental and exchange particles, which allowed for the establishment of a "standard model" based on the mathematics of gauge invariance, which successfully described all forces except for gravitation, and which remains generally accepted within its domain of application.

The Standard Model, based on the Yang-Mills Theory groups the electroweak interaction theory and quantum chromodynamics into a structure denoted by the gauge group SU(3)×SU(2)×U(1). The formulation of the unification of the electromagnetic and weak interactions in the standard model is due to Abdus Salam, Steven Weinberg and, subsequently, Sheldon Glashow. Electroweak theory was later confirmed experimentally (by observation of neutral weak currents), and distinguished by the 1979 Nobel Prize in Physics.

Since the 1970s, fundamental particle physics has provided insights into early universe cosmology, particularly the Big Bang theory proposed as a consequence of Einstein's general theory of relativity. However, starting in the 1990s, astronomical observations have also provided new challenges, such as the need for new explanations of galactic stability ("dark matter") and the apparent acceleration in the expansion of the universe ("dark energy").

While accelerators have confirmed most aspects of the Standard Model by detecting expected particle interactions at various collision energies, no theory reconciling general relativity with the Standard Model has yet been found, although supersymmetry and string theory were believed by many theorists to be a promising avenue forward. The Large Hadron Collider, however, which began operating in 2008, has failed to find any evidence whatsoever that is supportive of supersymmetry and string theory.

Cosmology

Cosmology may be said to have become a serious research question with the publication of Einstein's General Theory of Relativity in 1915 although it did not enter the scientific mainstream until the period known as the "Golden age of general relativity".

About a decade later, in the midst of what was dubbed the "Great Debate", Hubble and Slipher discovered the expansion of universe in the 1920s measuring the redshifts of Doppler spectra from galactic nebulae. Using Einstein's general relativity, Lemaître and Gamow formulated what would become known as the big bang theory. A rival, called the steady state theory, was devised by Hoyle, Gold, Narlikar and Bondi.

Cosmic background radiation was verified in the 1960s by Penzias and Wilson, and this discovery favoured the big bang at the expense of the steady state scenario. Later work was by Smoot et al. (1989), among other contributors, using data from the Cosmic Background explorer (CoBE) and the Wilkinson Microwave Anisotropy Probe (WMAP) satellites that refined these observations. The 1980s (the same decade of the COBE measurements) also saw the proposal of inflation theory by Alan Guth.

Higgs boson

One possible signature of a Higgs boson from a simulated proton–proton collision. It decays almost immediately into two jets of hadrons and two electrons, visible as lines.

On July 4, 2012, physicists working at CERN's Large Hadron Collider announced that they had discovered a new subatomic particle greatly resembling the Higgs boson, a potential key to an understanding of why elementary particles have mass and indeed to the existence of diversity and life in the universe. For now, some physicists are calling it a "Higgslike" particle. Joe Incandela, of the University of California, Santa Barbara, said, "It's something that may, in the end, be one of the biggest observations of any new phenomena in our field in the last 30 or 40 years, going way back to the discovery of quarks, for example." Michael Turner, a cosmologist at the University of Chicago and the chairman of the physics center board, said:

"This is a big moment for particle physics and a crossroads — will this be the high water mark or will it be the first of many discoveries that point us toward solving the really big questions that we have posed?"

—Michael Turner, University of Chicago

Peter Higgs was one of six physicists, working in three independent groups, who, in 1964, invented the notion of the Higgs field ("cosmic molasses"). The others were Tom Kibble of Imperial College, London; Carl Hagen of the University of Rochester; Gerald Guralnik of Brown University; and François Englert and Robert Brout, both of Université libre de Bruxelles.

Although they have never been seen, Higgslike fields play an important role in theories of the universe and in string theory. Under certain conditions, according to the strange accounting of Einsteinian physics, they can become suffused with energy that exerts an antigravitational force. Such fields have been proposed as the source of an enormous burst of expansion, known as inflation, early in the universe and, possibly, as the secret of the dark energy that now seems to be speeding up the expansion of the universe.

Physical Sciences

With increased accessibility to and elaboration upon advanced analytical techniques in the 19th century, physics was defined as much, if not more, by those techniques than by the search for universal principles of motion and energy, and the fundamental nature of matter. Fields such as acoustics, geophysics, astrophysics, aerodynamics, plasma physics, low-temperature physics, and solid-state physics joined optics, fluid dynamics, electromagnetism, and mechanics as areas of physical research. In the 20th century, physics also became closely allied with such fields as electrical, aerospace and materials engineering, and physicists began to work in government and industrial laboratories as much as in academic settings. Following World War II, the population of physicists increased dramatically, and came to be centered on the United States, while, in more recent decades, physics has become a more international pursuit than at any time in its previous history.

THE SCOPE OF PHYSICS

Mechanics is generally taken to mean the study of the motion of objects (or their lack of motion) under the action of given forces. Classical mechanics is sometimes considered a branch of applied mathematics. It consists of kinematics, the description of motion, and dynamics, the study of the action of forces in producing either motion or static equilibrium (the latter constituting the science of statics). The 20th-century subjects of quantum mechanics, crucial to treating the structure of matter, subatomic particles, superfluidity, superconductivity, neutron stars, and other major phenomena, and relativistic mechanics, important when speeds approach that of light, are forms of mechanics that will be discussed later in this section.

In classical mechanics the laws are initially formulated for point particles in which the dimensions, shapes, and other intrinsic properties of bodies are ignored. Thus in the first approximation even objects as large as Earth and the Sun are treated as pointlike—e.g., in calculating planetary orbital motion. In rigid-body dynamics, the extension of bodies and their mass distributions are considered as well, but they are imagined to be incapable of deformation. The mechanics of deformable solids is elasticity; hydrostatics and hydrodynamics treat, respectively, fluids at rest and in motion.

The three laws of motion set forth by Isaac Newton form the foundation of classical mechanics, together with the recognition that forces are directed quantities (vectors) and combine accordingly. The first law, also called the law of inertia, states that, unless acted upon by an external force, an object at rest remains at rest, or if in motion, it continues to move in a straight line with constant speed. Uniform motion therefore does not require a cause. Accordingly, mechanics concentrates not on motion as such but on the change in the state of motion of an object that results from the net force acting upon it. Newton's second law equates the net force on an object to the rate of change of its momentum, the latter being the product of the mass of a body and its velocity. Newton's third law, that of action and reaction, states that when two particles interact, the forces each exerts on the other are equal in magnitude and opposite in direction. Taken together, these mechanical laws in principle permit the determination of the future motions of a set of particles, providing their state of motion is known at some instant, as well as the forces that act between them and upon them from the outside. From this deterministic character of the laws of classical mechanics, profound (and probably incorrect) philosophical conclusions have been drawn in the past and even applied to human history.

Lying at the most basic level of physics, the laws of mechanics are characterized by certain symmetry properties, as exemplified in the aforementioned symmetry between action and reaction forces. Other symmetries, such as the invariance (i.e., unchanging form) of the laws under reflections and rotations carried out in space, reversal of time, or transformation to a different part of space or to a different epoch of time, are present both in classical mechanics and in relativistic mechanics, and with certain restrictions, also in quantum mechanics. The symmetry properties of the theory can be shown to have as mathematical consequences basic principles known as conservation laws, which assert the constancy in time of the values of certain physical quantities under prescribed conditions. The conserved quantities are the most important ones in physics; included among them are mass and energy (in relativity theory, mass and energy are equivalent and are conserved together), momentum, angular momentum, and electric charge.

CLASSICAL MECHANICS

Classical mechanics describes the motion of macroscopic objects, from projectiles to parts of machinery, and astronomical objects, such as spacecraft, planets, stars and galaxies. If the present state of an object is known it is possible to predict by the laws of classical mechanics how it will move in the future (determinism) and how it has moved in the past (reversibility). The earliest development of classical mechanics is often referred to as Newtonian mechanics. It consists of the physical concepts employed and the mathematical methods invented by Isaac Newton, Gottfried Wilhelm Leibniz and others in the 17th century to describe the motion of bodies under the influence of a system of forces.

Later, more abstract methods were developed, leading to the reformulations of classical mechanics known as Lagrangian mechanics and Hamiltonian mechanics. These advances, made predominantly in the 18th and 19th centuries, extend substantially beyond Newton's work, particularly through their use of analytical mechanics. They are, with some modification, also used in all areas of modern physics.

Classical mechanics provides extremely accurate results when studying large objects that are not extremely massive and speeds not approaching the speed of light. When the objects being examined have about the size of an atom diameter, it becomes necessary to introduce the other major sub-field of mechanics: quantum mechanics. To describe velocities that are not small compared to the speed of light, special relativity is needed. In case that objects become extremely massive, general relativity becomes applicable. However, a number of modern sources do include relativistic mechanics into classical physics, which in their view represents classical mechanics in its most developed and accurate form.

Wave

In physics, mathematics, and related fields, a wave is a disturbance (change from equilibrium) of one or more fields such that the field values oscillate repeatedly about a stable equilibrium (resting) value. If the relative amplitude of oscillation at different points in the field remains constant, the wave is said to be a standing wave. If the relative amplitude at different points in the field changes, the wave is said to be a traveling wave. Waves can only exist in fields when there is a force that tends to restore the field to equilibrium. The types of waves most commonly studied in physics are mechanical and electromagnetic. In a mechanical wave, stress and strain fields oscillate about a mechanical equilibrium. A traveling mechanical wave is a local deformation (strain) in some physical medium that propagates from particle to particle by creating local stresses that cause strain in neighboring particles too. For example, sound waves in air are variations of the local pressure that propagate by collisions between gas molecules. Other examples of mechanical waves are seismic waves, gravity waves, vortices, and shock waves. In an electromagnetic wave the electric and magnetic fields oscillate. A traveling electromagnetic wave (light) consists of a combination of variable electric and magnetic fields, that propagates through space according to Maxwell's equations. Electromagnetic waves can travel through transparent dielectric media or through a vacuum; examples include radio waves, infrared radiation, visible light, ultraviolet radiation, X-rays and gamma rays.

Other types of waves include gravitational waves, which are disturbances in a gravitational field that propagate according to general relativity; heat diffusion waves; plasma waves, that combine mechanical deformations and electromagnetic fields; reaction-diffusion waves, such as in the Belousov–Zhabotinsky reaction; and many more. Mechanical and electromagnetic waves transfer energy,, momentum, and information, but they do not transfer particles in the medium. In mathematics and electronics waves are studied as signals. On the other hand, some waves do not appear to move at all, like standing waves (which are fundamental to music) and hydraulic jumps. Some, like the probability waves of quantum mechanics, may be completely static.

A physical wave is almost always confined to some finite region of space, called its domain. For example, the seismic waves generated by earthquakes are significant only in the interior and surface of the planet, so they can be ignored outside it. However, waves with infinite domain, that extend over the whole space, are commonly studied in mathematics, and are very valuable tools for understanding physical waves in finite domains. A plane wave seems to travel in a definite direction, and has constant value over any plane perpendicular to that direction. Mathematically, the simplest waves are the sinusoidal ones in which each point in the field experiences simple harmonic motion. Complicated waves can often be described as the sum of many sinusoidal plane waves. A plane wave can be a transverse, if its effect at each point is described by a vector that is perpendicular to the direction of propagation or energy transfer; or longitudinal, if the describing vectors are parallel to the direction of energy propagation. While mechanical waves can be both transverse and longitudinal, electromagnetic waves are transverse in free space.

Thermodynamics

Thermodynamics is the branch of physics that deals with heat and temperature, and their relation to energy, work, radiation, and properties of matter. The behavior of these quantities is governed by the four laws of thermodynamics which convey a quantitative description using measurable macroscopic physical quantities, but may be explained in terms of microscopic constituents by statistical mechanics. Thermodynamics applies to a wide variety of topics in science and engineering, especially physical chemistry, chemical engineering and mechanical engineering, but also in fields as complex as meteorology.

Historically, thermodynamics developed out of a desire to increase the efficiency of early steam engines, particularly through the work of French physicist Nicolas Léonard Sadi Carnot (1824) who believed that engine efficiency was the key that could help France win the Napoleonic Wars. Scots-Irish physicist Lord Kelvin was the first to formulate a concise definition of thermodynamics in 1854 which stated, "Thermo-dynamics is the subject of the relation of heat to forces acting between contiguous parts of bodies, and the relation of heat to electrical agency."

The initial application of thermodynamics to mechanical heat engines was quickly extended to the study of chemical compounds and chemical reactions. Chemical thermodynamics studies the nature of the role of entropy in the process of chemical reactions and has provided the bulk of expansion and knowledge of the field. Other formulations of thermodynamics emerged. Statistical thermodynamics, or statistical mechanics, concerns itself with statistical predictions of the collective motion of particles from their microscopic behavior. In 1909, Constantin Carathéodory presented a purely mathematical approach in an axiomatic formulation, a description often referred to as geometrical thermodynamics.

Electromagnetism

Electromagnetism is a branch of physics involving the study of the electromagnetic force, a type of physical interaction that occurs between electrically charged particles. The electromagnetic force is carried by electromagnetic fields composed of electric fields and magnetic fields, and it is responsible for electromagnetic radiation such as light. It is one of the four fundamental interactions (commonly called forces) in nature, together with the strong interaction, the weak interaction, and gravitation. At high energy the weak force and electromagnetic force are unified as a single electroweak force.

Electromagnetic phenomena are defined in terms of the electromagnetic force, sometimes called the Lorentz force, which includes both electricity and magnetism as different manifestations of the same phenomenon. The electromagnetic force plays a major role in determining the internal properties of most objects encountered in daily life. The electromagnetic attraction between atomic nuclei and their orbital electrons holds atoms together. Electromagnetic forces are responsible for the chemical bonds between atoms which create molecules, and intermolecular forces. The electromagnetic force governs all chemical processes, which arise from interactions between the electrons of neighboring atoms.

There are numerous mathematical descriptions of the electromagnetic field. In classical electrodynamics, electric fields are described as electric potential and electric current. In Faraday's law, magnetic fields are associated with electromagnetic induction and magnetism, and Maxwell's equations describe how electric and magnetic fields are generated and altered by each other and by charges and currents. The theoretical implications of electromagnetism, particularly the establishment of the speed of light based on properties of the "medium" of propagation (permeability and permittivity), led to the development of special relativity by Albert Einstein in 1905.

Optics

Optics is the branch of physics that studies the behaviour and properties of light, including its interactions with matter and the construction of instruments that use or detect it. Optics usually describes the behaviour of visible, ultraviolet, and infrared light. Because light is an electromagnetic wave, other forms of electromagnetic radiation such as X-rays, microwaves, and radio waves exhibit similar properties.

Most optical phenomena can be accounted for using the classical electromagnetic description of light. Complete electromagnetic descriptions of light are, however, often difficult to apply in practice. Practical optics is usually done using simplified models. The most common of these, geometric optics, treats light as a collection of rays that travel in straight lines and bend when they pass through or reflect from surfaces. Physical optics is a more comprehensive model of light, which includes wave effects such as diffraction and interference that cannot be accounted for in geometric optics. Historically, the ray-based model of light was developed first, followed by the wave model of light. Progress in electromagnetic theory in the 19th century led to the discovery that light waves were in fact electromagnetic radiation. Some phenomena depend on the fact that light has both wave-like and particle-like properties. Explanation of these effects requires quantum mechanics. When considering light's particle-like properties, the light is modelled as a collection of particles called "photons". Quantum optics deals with the application of quantum mechanics to optical systems.

Optical science is relevant to and studied in many related disciplines including astronomy, various engineering fields, photography, and medicine (particularly ophthalmology and optometry). Practical applications of optics are found in a variety of technologies and everyday objects, including mirrors, lenses, telescopes, microscopes, lasers, and fibre optics. The extended edition also contains introductions to topics such as quantum mechanics, atomic theory, solid-state physics, nuclear physics and cosmology. A solutions manual and a study guide are also available.

2

MECHANICS

Kinematics is a branch of classical mechanics that describes the motion of points, bodies (objects), and systems of bodies (groups of objects) without considering the forces that cause them to move. Kinematics, as a field of study, is often referred to as the "geometry of motion" and is occasionally seen as a branch of mathematics. A kinematics problem begins by describing the geometry of the system and declaring the initial conditions of any known values of position, velocity and/or acceleration of points within the system.

Then, using arguments from geometry, the position, velocity and acceleration of any unknown parts of the system can be determined. The study of how forces act on bodies falls within kinetics, not kinematics. For further details, see analytical dynamics.

Kinematics is used in astrophysics to describe the motion of celestial bodies and collections of such bodies. In mechanical engineering, robotics, and biomechanics kinematics is used to describe the motion of systems composed of joined parts (multi-link systems) such as an engine, a robotic arm or the human skeleton. Geometric transformations, also called rigid transformations, are used to describe the movement of components in a mechanical system, simplifying the derivation of the equations of motion. They are also central to dynamic analysis. Kinematic analysis is the process of measuring the kinematic quantities used to describe motion. In engineering, for instance, kinematic analysis may be used to find the range of movement for a given mechanism and working in reverse, using kinematic synthesis to design a mechanism for a desired range of motion. In addition, kinematics applies algebraic geometry to the study of the mechanical advantage of a mechanical system or mechanism.

ETYMOLOGY OF THE TERM

The term kinematic is the English version of A.M. Ampère's cinématique, which he constructed from the Greek kinema ("movement, motion"), itself derived from kinein ("to move"). Kinematic and cinématique are related to the French word cinéma, but neither are directly derived from it. However, they do share a root word in common, as cinéma came from the shortened form of cinématographe, "motion picture projector and camera," once again from the Greek word for movement and from the Greek grapho ("to write").

Velocity and speed

The velocity of a particle is a vector quantity that describes the magnitude as well as direction of motion of the particle. More mathematically, the rate of change of the position vector of a point, with respect to time is the velocity of the point. Consider the ratio formed by dividing the difference of two positions of a particle by the time interval. This ratio is called the average velocity over that time interval and is defined as Velocity=displacement/time taken

$$\bar{\mathbf{V}} = \frac{\Delta \mathbf{P}}{\Delta t},$$

where ÄP is the change in the position vector over the time interval Ät. In the limit as the time interval Ät becomes smaller and smaller, the average velocity becomes the time derivative of the position vector,

$$\mathbf{V} = \lim_{\Delta t \to 0} \frac{\Delta \mathbf{P}}{\Delta t} = \frac{d\mathbf{P}}{dt} = \dot{\mathbf{P}} = \dot{x}_p \hat{i} + \dot{y}_p \hat{j} + \dot{z}_p \hat{k}.$$

Thus, velocity is the time rate of change of position of a point, and the dot denotes the derivative of those functions x, y, and z with respect to time. Furthermore, the velocity is tangent to the trajectory of the particle at every position the particle occupies along its path. Note that in a non-rotating frame of reference, the derivatives of the coordinate directions are not considered as their directions and magnitudes are constants. The speed of an object is the magnitude |V| of its velocity. It is a scalar quantity:

$$|\mathbf{V}| = |\dot{\mathbf{P}}| = \frac{ds}{dt},$$

wheres is the arc-length measured along the trajectory of the particle. This arc-length traveled by a particle over time is a non-decreasing quantity. Hence, ds/dt is non-negative, which implies that speed is also non-negative.

Acceleration

The velocity vector can change in magnitude and in direction or both at once. Hence, the acceleration accounts for both the rate of change of the magnitude of the velocity vector and the rate of change of direction of that vector. The same reasoning used with respect to the position of a particle to define velocity, can be applied to the velocity to define acceleration. The acceleration of a particle is the vector defined by the rate of change of the velocity vector. The average acceleration of a particle over a time interval is defined as the ratio.

$$\bar{\mathbf{A}} = \frac{\Delta \mathbf{V}}{\Delta t},$$

where ÄV is the difference in the velocity vector and ?t is the time interval.

The acceleration of the particle is the limit of the average acceleration as the time interval approaches zero, which is the time derivative,

$$\mathbf{A} = \lim_{\Delta t \to 0} \frac{\Delta \mathbf{V}}{\Delta t} = \frac{d\mathbf{V}}{dt} = \dot{\mathbf{V}} = \dot{v}_x \hat{i} + \dot{v}_y \hat{j} + \dot{v}_z \hat{k}$$

or

$$\mathbf{A} = \ddot{\mathbf{P}} = \ddot{x}_p \hat{i} + \ddot{y}_p \hat{j} + \ddot{z}_p \hat{k}$$

Thus, acceleration is the first derivative of the velocity vector and the second derivative of the position vector of that particle. Note that in a non-rotating frame of reference, the derivatives of the coordinate directions are not considered as their directions and magnitudes are constants.

The magnitude of the acceleration of an object is the magnitude |A| of its acceleration vector. It is a scalar quantity:

$$|\mathbf{A}| = |\dot{\mathbf{V}}| = \frac{dv}{dt},$$

Relative position vector

A relative position vector is a vector that defines the position of one point relative to another. It is the difference in position of the two points. The position of one point A relative to another point B is simply the difference between their positions

$$P_{A/B} = P_A = P_B$$

which is the difference between the components of their position vectors.

If point A has position components

$$P_A = (X_A, Y_A, Z_A)$$

If point B has position components P B = (X B, Y B, Z B) {\displaystyle \mathbf {P} _{B}=\left(X_{B},Y_{B},Z_{B}\right)}

$$P_B = (X_B, Y_B, Z_B)$$

then the position of point A relative to point B is the difference between their components:

$$P_{A/B} = P_A - P_B = (X_A - X_B, Y_A - Y_B, Z_A - Z_B)$$

Relative velocity

The velocity of one point relative to another is simply the difference between their velocities

$$V_{A/B} = V_A - V_B$$

which is the difference between the components of their velocities. If point A has velocity components

$$V_A = (A_{A_x}, V_{A_y}, V_{A_Z})$$

and point B has velocity component

$$VB = (V_{B_x}, V_{B_y}, V_{B_z},)$$

then the velocity of point A relative to point B is the difference between their components:

$$V_{A/B} = V_A - V_B = (V_{A_x} - V_{B_x}, V_{A_y} - V_{B_y}, V_{A_z}, - V_{B_z},)$$

Alternatively, this same result could be obtained by computing the time derivative of the relative position vector R. In the case where the velocity is close to the speed of lightc (generally within 95%), another scheme of relative velocity called rapidity, that depends on the ratio of V to c, is used in special relativity.

Point trajectories in a body moving in the plane

The movement of components of a mechanical system are analyzed by attaching a reference frame to each part and determining how the various reference frames move relative to each other. If the structural stiffness of the parts are sufficient, then their deformation can be neglected and rigid transformations can be used to define this relative movement. This reduces the description of the motion of the various parts of a complicated mechanical system to a problem of describing the geometry of each part and geometric association of each part relative to other parts.

Geometry is the study of the properties of figures that remain the same while the space is transformed in various ways-more technically, it is the study of invariants under a set of transformations. These transformations can cause the displacement of the triangle in the plane, while leaving the vertex angle and the distances between vertices unchanged. Kinematics is often described as applied geometry, where the movement of a mechanical system is described using the rigid transformations of Euclidean geometry.

The coordinates of points in a plane are two-dimensional vectors in R (two dimensional space). Rigid transformations are those that preserve the distance between any two points. The set of rigid transformations in an n-dimensional space is called the special Euclidean group on R, and denoted SE(n).

Displacements and motion

The position of one component of a mechanical system relative to another is defined by introducing a reference frame, say M, on one that moves relative to a fixed frame, F, on the other. The rigid transformation, or displacement, of M relative to F defines the relative position of the two components. A displacement consists of the combination of a rotation and a translation.

The set of all displacements of M relative to F is called the configuration space of M. A smooth curve from one position to another in this configuration space is a continuous set of displacements, called the motion of M relative to F. The motion of a body consists of a continuous set of rotations and translations.

Projectile motion

Projectile motion is the motion of an object thrown, or projected, into the air, subject only to the force of gravity. The object is called a projectile, and its path is called its trajectory. The motion of falling objects is a simple one-dimensional type of projectile motion in which there is no horizontal movement. In two-dimensional projectile motion, such as that of a football or other thrown object, there is both a vertical and a horizontal component to the motion.

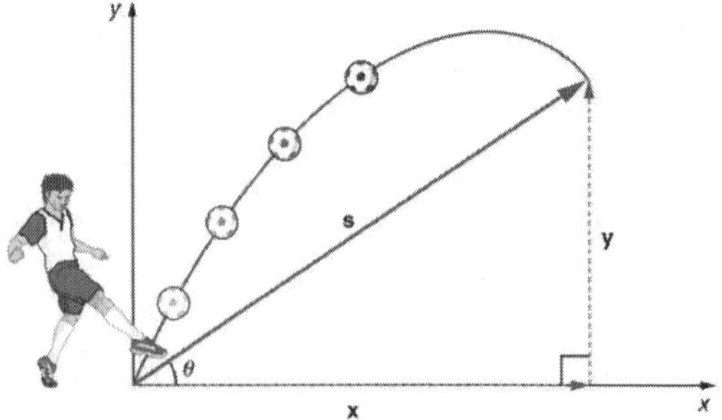

Projectile Motion: Throwing a rock or kicking a ball generally produces a projectile pattern of motion that has both a vertical and a horizontal component. The most important fact to remember is that motion along perpendicular axes are independent and thus can be analyzed separately. The key to analyzing two-dimensional projectile motion is to break it into two motions, one along the horizontal axis and the other along the vertical. To describe motion we must deal with velocity and acceleration, as well as with displacement.

We will assume all forces except for gravity (such as air resistance and friction, for example) are negligible. The components of acceleration are then very simple: ay="g="9.81ms2 (we assume that the motion occurs at small enough heights near the surface of the earth so that the acceleration due to gravity is constant). Because the acceleration due to gravity is along the vertical direction only, ax=0. Thus, the kinematic equations describing the motion along the x and y directions respectively, can be used:

x=x0+vxt

vy=v0y+ayt

y=y0+v0yt+12ayt2

v2y=v20y+2ay(y"y0)

Projectile motion is a form of motion experienced by an object or particle (a projectile) that is thrown near the Earth's surface and m•oves along a curved path under the action of gravity only (in particular, the effects of air resistance are assumed to be negligible). This curved path was shown by Galileo to be a parabola. The study of such motions is called ballistics, and such a trajectory is a ballistic trajectory. The only force of significance that acts on the object is gravity, which acts downward, thus imparting to the object a downward acceleration. Because of the object's inertia, no external horizontal force is needed to maintain the horizontal velocity component of the object. Taking other forces into account, such as friction from aerodynamic drag or internal propulsion such as in a rocket, requires additional analysis. A ballistic missile is a missile only guided during the relatively brief initial powered phase of flight, and whose subsequent course is governed by the laws of classical mechanics.

Ballistics (gr. âÜëëåéí ('ba'llein'), "to throw") is the science of mechanics that deals with the flight, behavior, and effects of projectiles, especially bullets, unguided bombs, rockets, or the like; the science or art of designing and accelerating projectiles so as to achieve a desired performance.

The elementary equations of ballistics neglect nearly every factor except for initial velocity and an assumed constant gravitational acceleration. Practical solutions of a ballistics problem often require considerations of air resistance, cross winds, target motion, varying acceleration due to gravity, and in such problems as launching a rocket from one point on the Earth to another, the rotation of the Earth. Detailed mathematical solutions of practical problems typically do not have closed-form solutions, and therefore require numerical methods to address.

We analyze two-dimensional projectile motion by breaking it into two independent one-dimensional motions along the vertical and horizontal axes. The horizontal motion is simple, because ax=0 and vx is thus constant. The velocity in the vertical direction begins to decrease as an object rises; at its highest point, the vertical velocity is zero. As an object falls towards the Earth again, the vertical velocity increases again in magnitude but points in the opposite direction to the initial vertical velocity. The x and y motions can be recombined to give the total velocity at any given point on the trajectory.

The initial velocity

Let the projectile be launched with an initial velocity,

$$v(0) \equiv v_0$$

which can be expressed as the sum of horizontal and vertical components as follows:

$$v_0 = v_{0x} i + v_{0y} j$$

The components v 0 x {displaystyle v_{0x}}and v 0 y {\displaystyle v_{0y}}

can be found if the initial launch angle,is known:

$$v_{0x} = v_0 \cos \theta$$

$$v_{0y} = v_0 \sin \theta$$

Projectile.A body which is in flight through the atmosphere but is not being propelled by any fuel is called projectile. Example: (i) A bomb released from an aeroplane in level flight (ii) A bullet fired from a gun(iii) An arrow released from bow (iv) A Javelin thrown by an athlete

Assumptions of Projectile Motion

(1) There is no resistance due to air.

(2) The effect due to curvature of earth is negligible.

(3) The effect due to rotation of earth is negligible.

(4) For all points of the trajectory, the acceleration due to gravity 'g' is constant in magnitude and direction.

Principles of Physical Independence of Motions.

(1) The motion of a projectile is a two-dimensional motion. So, it can be discussed in two parts. Horizontal motion and vertical motion. These two motions take place independent of each other. This is called the principle of physical independence of motions.

(2) The velocity of the particle can be resolved into two mutually perpendicular components. Horizontal component and vertical component.

(3) The horizontal component remains unchanged throughout the flight. The force of gravity continuously affects the vertical component.

(4) The horizontal motion is a uniform motion and the vertical motion is a uniformly accelerated retarded motion

Newton's laws of motion

Newton's laws of motion are three physical laws that, together, laid the foundation for classical mechanics. They describe the relationship between a body and the forces acting upon it, and its motion in response to those forces. More precisely, the first law defines the force qualitatively, the second law offers a quantitative measure of the force, and the third asserts that a single isolated force doesn't exist. These three laws have been expressed in several ways, over nearly three centuries, and can be summarised as follows:

First law: In an inertial frame of reference, an object either remains at rest or continues to move at a constant velocity, unless acted upon by a force.

Second law: In an inertial frame of reference, the vector sum of the forcesF on an object is equal to the massm of that object multiplied by the accelerationa of the object: F = ma. (It is assumed here that the mass m is constant - see below.)

Third law: When one body exerts a force on a second body, the second body simultaneously exerts a force equal in magnitude and opposite in direction on the first body. The three laws of motion were first compiled by Isaac Newton in his Philosophiæ Naturalis Principia Mathematica (Mathematical Principles of Natural Philosophy), first published in 1687. Newton used them to explain and investigate the motion of many physical objects and systems. For example, in the third volume of the text, Newton showed that these laws of motion, combined with his law of universal gravitation, explained Kepler's laws of planetary motion. Some also describe a fourth law which states that forces add up like vectors, that is, that forces obey the principle of superposition.

Overview

Newton's laws are applied to objects which are idealised as single point masses, in the sense that the size and shape of the object's body are neglected to focus on its motion more easily. This can be done when the object is small compared to the distances involved in its analysis, or the deformation and rotation of the body are of no importance. In this way, even a planet can be idealised as a particle for analysis of its orbital motion around a star.

In their original form, Newton's laws of motion are not adequate to characterise the motion of rigid bodies and deformable bodies. Leonhard Euler in 1750 introduced a generalisation of Newton's laws of motion for rigid bodies called Euler's laws of motion, later applied as well for deformable bodies assumed as a continuum. If a body is represented as an assemblage of discrete particles, each governed by Newton's laws of motion, then Euler's laws can be derived from Newton's laws. Euler's laws can, however, be taken as axioms describing the laws of motion for extended bodies, independently of any particle structure.

Newton's laws hold only with respect to a certain set of frames of reference called Newtonian or inertial reference frames. Some authors interpret the first law as defining what an inertial reference frame is; from this point of view, the second law holds only when the observation is made from an inertial reference frame, and therefore the first law cannot be proved as a special case of the second. Other authors do treat the first law as a corollary of the second. The explicit concept of an inertial frame of reference was not developed until long after Newton's death. In the given interpretation mass, acceleration, momentum, and (most importantly) force are assumed to be externally defined quantities. This is the most common, but not the only interpretation of the way one can consider the laws to be a definition of these quantities. Newtonian mechanics has been superseded by special relativity, but it is still useful as an approximation when the speeds involved are much slower than the speed of light.

Laws

Newton's laws read:

Law I: Every body persists in its state of being at rest or of moving uniformly straight forward, except insofar as it is compelled to change its state by force impressed.

Law II: The alteration of motion is ever proportional to the motive force impress'd; and is made in the direction of the right line in which that force is impress'd.

Law III: To every action there is always opposed an equal reaction: or the mutual actions of two bodies upon each other are always equal, and directed to contrary parts.

Newton's first law

The first law states that if the net force (the vector sum of all forces acting on an object) is zero, then the velocity of the object is constant. Velocity is a vector quantity which expresses both the object's speed and the direction of its motion; therefore, the statement that the object's velocity is constant is a statement that both its speed and the direction of its motion are constant.

The first law can be stated mathematically when the mass is a non-zero constant, as,

$$\sum \mathbf{F} = 0 \Leftrightarrow \frac{\mathrm{d}\mathbf{v}}{\mathrm{d}t} = 0.$$

Consequently,

- An object that is at rest will stay at rest unless a force acts upon it.
- An object that is in motion will not change its velocity unless a force acts upon it.

This is known as uniform motion. An object continues to do whatever it happens to be doing unless a force is exerted upon it. If it is at rest, it continues in a state of rest (demonstrated when a tablecloth is skilfully whipped from under dishes on a tabletop and the dishes remain in their initial state of rest). If an object is moving, it continues to move without turning or changing its speed. This is evident in space probes that continuously move in outer space. Changes in motion must be imposed against the tendency of an object to retain its state of motion. In the absence of net forces, a moving object tends to move along a straight line path indefinitely.

Newton placed the first law of motion to establish frames of reference for which the other laws are applicable. However, Newton implicitly referred to the absolute co-ordinate of cosmos for this frame. Since we cannot precisely measure our velocity relative to a far star, Newton's frame is based on a pure imagination, not based on measurable physics. In current physics, an observer defines himself as in inertial frame by preparing one stone hooked by a spring, and rotating the spring to any direction, and observing the stone static and the length of that spring unchanged. By Einstein's equivalence principle, if there was one such observer A and another observer B moving in a constant velocity related to A, then A and B will both observe the same physics phenomena. if A verified the first law, then

B will verify it too. In this way, the definition of inertial can get rid of absolute space or far star, and only refer to the objects locally reachable and measurable. A particle not subject to forces moves (related to inertial frame) in a straight line at a constant speed. Newton's first law is often referred to as the law of inertia. Thus, a condition necessary for the uniform motion of a particle relative to an inertial reference frame is that the total net force acting on it is zero. In this sense, the first law can be restated as:

In every material universe, the motion of a particle in a preferential reference frame Ö is determined by the action of forces whose total vanished for all times when and only when the velocity of the particle is constant in Ö. That is, a particle initially at rest or in uniform motion in the preferential frame Ö continues in that state unless compelled by forces to change it. Newton's first and second laws are valid only in an inertial reference frame. Any reference frame that is in uniform motion with respect to an inertial frame is also an inertial frame, i.e. Galilean invariance or the principle of Newtonian relativity.

Newton's second law

The second law states that the rate of change of momentum of a body is directly proportional to the force applied, and this change in momentum takes place in the direction of the applied force.

$$\mathbf{F} = \frac{\mathrm{d}\mathbf{p}}{\mathrm{d}t} = \frac{\mathrm{d}(m\mathbf{v})}{\mathrm{d}t}.$$

The second law can also be stated in terms of an object's acceleration. Since Newton's second law is valid only for constant-mass systems,m can be taken outside the differentiation operator by the constant factor rule in differentiation. Thus,

$$\mathbf{F} = m\frac{\mathrm{d}\mathbf{v}}{\mathrm{d}t} = m\mathbf{a},$$

where F is the net force applied, m is the mass of the body, and a is the body's acceleration. Thus, the net force applied to a body produces a proportional acceleration. In other words, if a body is accelerating, then there is a force on it. An application of this notation is the derivation of G Subscript C.

The above statements hint that the second law is merely a definition of F, not a precious observation of nature. However, current physics restate the second law in measurable steps: (1)defining the term 'one unit of mass' by a specified stone, (2)defining the term 'one unit of force' by a specified spring with specified length, (3)measuring by experiment or proving by theory (with a principle that every direction of space are equivalent), that force can be added as a mathematical vector, (4)finally conclude that F = m a These steps hint the second law is a precious feature of nature.

The second law also implies the conservation of momentum: when the net force on the body is zero, the momentum of the body is constant. Any net force is equal to the rate of change of the momentum. Any mass that is gained or lost by the system will cause a change in momentum that is not the result of an external force. A different equation is necessary for variable-mass systems (see below). Newton's second law is an approximation that is increasingly worse at high speeds because of relativistic effects. According to modern ideas of how Newton was using his terminology, the law is understood, in modern terms, as an equivalent of:

The change of momentum of a body is proportional to the impulse impressed on the body, and happens along the straight line on which that impulse is impressed. This may be expressed by the formula F = p', where p' is the time derivative of the momentum p. This equation can be seen clearly in the Wren Library of Trinity College, Cambridge, in a glass case in which Newton's manuscript is open to the relevant page.

Motte's 1729 translation of Newton's Latin continued with Newton's commentary on the second law of motion, reading: If a force generates a motion, a double force will generate double the motion, a triple force triple the motion, whether that force be impressed altogether and at once, or gradually and successively. And this motion (being always directed the same way with the generating force), if the body moved before, is added to or subtracted from the former motion, according as they directly conspire with or are directly contrary to each other; or obliquely joined, when they are oblique, so as to produce a new motion compounded from the determination of both. The sense or senses in which Newton used his terminology, and how he understood the second law and intended it to be understood, have been extensively discussed by historians of science, along with the relations between Newton's formulation and modern formulations.

Impulse

An impulse J occurs when a force F acts over an interval of time Ät, and it is given by

$$\mathbf{J} = \int_{\Delta t} \mathbf{F}\,\mathrm{d}t.$$

Since force is the time derivative of momentum, it follows that

$$J = \Delta p = m\Delta v.$$

This relation between impulse and momentum is closer to Newton's wording of the second law. Impulse is a concept frequently used in the analysis of collisions and impacts.

Variable-mass systems

Variable-mass systems, like a rocket burning fuel and ejecting spent gases, are not closed and cannot be directly treated by making mass a function of time in the second law; that is, the following formula is wrong:

$$\mathbf{F}_{\text{net}} = \frac{\mathrm{d}}{\mathrm{d}t}[m(t)\mathbf{v}(t)] = m(t)\frac{\mathrm{d}\mathbf{v}}{\mathrm{d}t} + \mathbf{v}(t)\frac{\mathrm{d}m}{\mathrm{d}t}. \qquad \text{(wrong)}$$

The falsehood of this formula can be seen by noting that it does not respect Galilean invariance: a variable-mass object with F = 0 in one frame will be seen to have F ‘d 0 in another frame. The correct equation of motion for a body whose mass m varies with time by either ejecting or accreting mass is obtained by applying the second law to the entire, constant-mass system consisting of the body and its ejected/accreted mass; the result iswhereu is the velocity of the escaping or incoming mass relative to the body. From this equation one can derive the equation of motion for a varying mass system, for example, the Tsiolkovsky rocket equation. Under some conventions, the quantity u dm/dt on the left-hand side, which represents the advection of momentum, is defined as a force (the force exerted on the body by the changing mass, such as rocket exhaust) and is included in the quantity F. Then, by substituting the definition of acceleration, the equation becomes F = ma.

Newton's third law

The third law states that all forces between two objects exist in equal magnitude and opposite direction: if one object A exerts a force F on a second object B, then B simultaneously exerts a force F on A, and the two forces are equal in magnitude and opposite in direction: F = “F. The third law means that all forces are interactions between different bodies, or different regions within one body, and thus that there is no such thing as a force that is not accompanied by an equal and opposite force. In some situations, the magnitude and direction of the forces are determined entirely by one of the two bodies, say Body A; the force exerted by Body A on Body B is called the “action”, and the force exerted by Body B on Body A is called the “reaction”. This law is sometimes referred to as the action-reaction law, with F called the “action” and F the “reaction”. In other situations the magnitude and directions of the forces are determined jointly by both bodies and it isn't necessary to identify one force as the "action" and the other as the "reaction". The action and the reaction are simultaneous, and it does not matter which is called the action and which is called reaction; both forces are part of a single interaction, and neither force exists without the other. The two forces in Newton's third law are of the same type (e.g., if the road exerts a forward frictional force on an accelerating car's tires, then it is also a frictional force that Newton's third law predicts for the tires pushing backward on the road).

From a conceptual standpoint, Newton's third law is seen when a person walks: they push against the floor, and the floor pushes against the person. Similarly, the tires of a car push against the road while the road pushes back on the tires—the tires and road simultaneously push against each other. In swimming, a person interacts with the water, pushing the water backward, while the water simultaneously pushes the person forward—both the person and the water push against each other. The reaction forces account for the motion in these examples. These forces depend on friction; a person or car on ice, for example, may be unable to exert the action force to produce the needed reaction force.

Newton used the third law to derive the law of conservation of momentum; from a deeper perspective, however, conservation of momentum is the more fundamental idea (derived via Noether's theorem from Galilean invariance), and holds in cases where Newton's third law appears to fail, for instance when force fields as well as particles carry momentum, and in quantum mechanics.

History

The ancient Greek philosopher Aristotle had the view that all objects have a natural place in the universe: that heavy objects (such as rocks) wanted to be at rest on the Earth and that light objects like smoke wanted to be at rest in the sky and the stars wanted to remain in the heavens. He thought that a body was in its natural state when it was at rest, and for the body to move in a straight line at a constant speed an external agent was needed continually to propel it, otherwise it would stop moving. Galileo Galilei, however, realised that a force is necessary to change the velocity of a body, i.e., acceleration, but no force is needed to maintain its velocity. In other words, Galileo stated that, in the absence of a force, a moving object will continue moving. (The tendency of objects to resist changes in motion was what Johannes Kepler had called inertia.) This insight was refined by Newton, who made it into his first law, also known as the "law of inertia"—no force means no acceleration, and hence the body will maintain its velocity. As Newton's first law is a restatement of the law of inertia which Galileo had already described, Newton appropriately gave credit to Galileo.

Leonardo da Vinci understood that "An object offers as much resistance to the air as the air does to the object". The law of inertia apparently occurred to several different natural philosophers and scientists independently, including Thomas Hobbes in his Leviathan (1651). The 17th-century philosopher and mathematician René Descartes also formulated the law, although he did not perform any experiments to confirm it.

Importance and range of validity

Newton's laws were verified by experiment and observation for over 200 years, and they are excellent approximations at the scales and speeds of everyday life. Newton's laws of motion, together with his law of universal gravitation and the mathematical techniques of calculus, provided for the first time a unified quantitative explanation for a wide range of physical phenomena.

These three laws hold to a good approximation for macroscopic objects under everyday conditions. However, Newton's laws (combined with universal gravitation and classical electrodynamics) are inappropriate for use in certain circumstances, most notably at very small scales, at very high speeds, or in very strong gravitational fields. Therefore, the laws cannot be used to explain phenomena such as conduction of electricity in a semiconductor, optical properties of substances, errors in non-relativistically corrected GPS systems and superconductivity. Explanation of these phenomena requires more sophisticated physical theories, including general relativity and quantum field theory.

In quantum mechanics, concepts such as force, momentum, and position are defined by linear operators that operate on the quantum state; at speeds that are much lower than the speed of light, Newton's laws are just as exact for these operators as they are for classical objects. At speeds comparable to the speed of light, the second law holds in the original form $F = dp/dt$, where F and p are four-vectors.

Relationship to the conservation laws

In modern physics, the laws of conservation of momentum, energy, and angular momentum are of more general validity than Newton's laws, since they apply to both light and matter, and to both classical and non-classical physics. This can be stated simply, "Momentum, energy and angular momentum cannot be created or destroyed." Because force is the time derivative of momentum, the concept of force is redundant and subordinate to the conservation of momentum, and is not used in fundamental theories (e.g., quantum mechanics, quantum electrodynamics, general relativity, etc.). The standard model explains in detail how the three fundamental forces known as gauge forces originate out of exchange by virtual particles. Other forces, such as gravity and fermionic degeneracy pressure, also arise from the momentum conservation. Indeed, the conservation of 4-momentum in inertial motion via curved space-time results in what we call gravitational force in general relativity theory. The application of the space derivative (which is a momentum operator in quantum mechanics) to the overlapping wave functions of a pair of fermions (particles with half-integer spin) results in shifts of maxima of compound wavefunction away from each other, which is observable as the "repulsion" of the fermions.

Newton stated the third law within a world-view that assumed instantaneous action at a distance between material particles. However, he was prepared for philosophical criticism of this action at a distance, and it was in this context that he stated the famous phrase "I feign no hypotheses". In modern physics, action at a distance has been completely eliminated, except for subtle effects involving quantum entanglement. (In particular, this refers to Bell's theorem—that no local model can reproduce the predictions of quantum theory.) Despite only being an approximation, in modern engineering and all practical applications involving the motion of vehicles and satellites, the concept of action at a distance is used extensively.

The discovery of the second law of thermodynamics by Carnot in the 19th century showed that not every physical quantity is conserved over time, thus disproving the validity of inducing the opposite metaphysical view from Newton's laws. Hence, a "steady-state" worldview based solely on Newton's laws and the conservation laws does not take entropy into account.

Momentum

In Newtonian mechanics, linear momentum, translational momentum, or simply momentum (pl. momenta) is the product of the mass and velocity of an object. It is a vector quantity, possessing a magnitude and a direction. If m is an object's mass and v is its velocity (also a vector quantity), then the object's momentum is:

$$p = m\,v.$$

In SI units, momentum is measured in kilogram meters per second (kgÅ"m/s). Newton's second law of motion states that a body's rate of change in momentum is equal to the net force acting on it. Momentum depends on the frame of reference, but in any inertial frame it is a conserved quantity, meaning that if a closed system is not affected by external forces, its total linear momentum does not change.

Momentum is also conserved in special relativity (with a modified formula) and, in a modified form, in electrodynamics, quantum mechanics, quantum field theory, and general relativity. It is an expression of one of the fundamental symmetries of space and time: translational symmetry.

Advanced formulations of classical mechanics, Lagrangian and Hamiltonian mechanics, allow one to choose coordinate systems that incorporate symmetries and constraints. In these systems the conserved quantity is generalized momentum, and in general this is different from the kinetic momentum defined above. The concept of generalized momentum is carried over into quantum mechanics, where it becomes an operator on a wave function. The momentum and position operators are related by the Heisenberg uncertainty principle.

In continuous systems such as electromagnetic fields, fluids and deformable bodies, a momentum density can be defined, and a continuum version of the conservation of momentum leads to equations such as the Navier–Stokes equations for fluids or the Cauchy momentum equation for deformable solids or fluids.

Conservation of momentum, general law of physics according to which the quantity called momentum that characterizes motion never changes in an isolated collection of objects; that is, the total momentum of a system remains constant. Momentum is equal to the mass of an object multiplied by its velocity and is equivalent to the force required to bring the object to a stop in a unit length of time. For any array of several objects, the total momentum is the sum of the individual momenta. There is a peculiarity, however, in that momentum is a vector, involving both the direction and the magnitude of motion, so that the momenta of objects going in opposite directions can cancel to yield an overall sum of zero.

Before launch, the total momentum of a rocket and its fuel is zero. During launch, the downward momentum of the expanding exhaust gases just equals in magnitude the upward momentum of the rising rocket, so that the total momentum of the system remains constant—in this case, at zero value. In a collision of two particles, the sum of the two momenta before collision is equal to their sum after collision. What momentum one particle loses, the other gains. The law of conservation of momentum is abundantly confirmed by experiment and can even be mathematically deduced on the reasonable presumption that space is uniform—that is, that there is nothing in the laws of nature that singles out one position in space as peculiar compared with any other.

There is a similar conservation law for angular momentum, which describes rotational motion in essentially the same way that ordinary momentum describes linear motion. Although the precise mathematical expression of this law is somewhat more involved, examples of it are numerous. All helicopters, for instance, require at least two propellers (rotors) for stabilization. The body of a helicopter would rotate in the opposite direction to conserve angular momentum if there were only a single horizontal propeller on top. In accordance with conservation of angular momentum, ice skaters spin faster as they pull their arms toward their body and more slowly as they extend them.

Circular motion

In physics, circular motion is a movement of an object along the circumference of a circle or rotation along a circular path. It can be uniform, with constant angular rate of rotation and constant speed, or non-uniform with a changing rate of rotation. The rotation around a fixed axis of a three-dimensional body involves circular motion of its parts. The equations of motion describe the movement of the center of mass of a body.

Examples of circular motion include: an artificial satellite orbiting the Earth at a constant height, a ceiling fan's blades rotating around a hub, a stone which is tied to a rope and is being swung in circles, a car turning through a curve in a race track, an electron moving perpendicular to a uniform magnetic field, and a gear turning inside a mechanism. Since the object's velocity vector is constantly changing direction, the moving object is undergoing acceleration by a centripetal force in the direction of the center of rotation. Without this acceleration, the object would move in a straight line, according to Newton's laws of motion.

Uniform circular motion

In physics, uniform circular motion describes the motion of a body traversing a circular path at constant speed. Since the body describes circular motion, its distance from the axis of rotation remains constant at all times. Though the body's speed is constant, its velocity is not constant: velocity, a vector quantity, depends on both the body's speed and its direction of travel. This changing velocity indicates the presence of an acceleration; this centripetal acceleration is of constant magnitude and directed at all times towards the axis of rotation. This acceleration is, in turn, produced by a centripetal force which is also constant in magnitude and directed towards the axis of rotation.

In the case of rotation around a fixed axis of a rigid body that is not negligibly small compared to the radius of the path, each particle of the body describes a uniform circular motion with the same angular velocity, but with velocity and acceleration varying with the position with respect to the axis

Centre of gravity

Centre of gravity, in physics, an imaginary point in a body of matter where, for convenience in certain calculations, the total weight of the body may be thought to be concentrated. The concept is sometimes useful in designing static structures (e.g., buildings and bridges) or in predicting the behaviour of a moving body when it is acted on by gravity.

In a uniform gravitational field the centre of gravity is identical to the centre of mass, a term preferred by physicists. The two do not always coincide, however. For example, the Moon's centre of mass is very close to its geometric centre (it is not exact because the Moon is not a perfect uniform sphere), but its centre of gravity is slightly displaced toward Earth because of the stronger gravitational force on the Moon's near side.

The location of a body's centre of gravity may coincide with the geometric centre of the body, especially in a symmetrically shaped object composed of homogeneous material. An asymmetrical object composed of a variety of materials with different masses, however, is likely to have a centre of gravity located at some distance from its geometric centre. In some cases, such as hollow bodies or irregularly shaped objects, the centre of gravity (or centre of mass) may occur in space at a point external to the physical material—e.g., in the centre of a tennis ball or between the legs of a chair.

Published tables and handbooks list the centres of gravity for most common geometric shapes. For a triangular metal plate such as that depicted in the figure, the calculation would involve a summation of the moments of the weights of all the particles that make up the metal plate about point A. By equating this sum to the plate's weight W, multiplied by the unknown distance from the centre of gravity G to AC, the position of G relative to AC can be determined. The summation of the moments can be obtained easily and precisely by means of integralcalculus

ROTATIONAL MOTION/ROTATIONAL MOTION

Old articles to be merged

In of linear motion, objects were treated as point particles without structure; it did not matter where a force was applied, only whether it was applied or not. The reality is that the point of application of force does matter. In tennis, for example, if a tennis ball is struck with a strong horizontal force acting through its center of mass, it may travel a long distance before hitting the ground, far out of bounds. Instead, the same force applied in an upward, glancing stoke will yield topspin to the ball, which can cause it to land in the opponent's court.

The concept of rotational equilibrium and rotational dynamics are also important in other disciplines. For example, students of architecture benefit from understanding the forces that act on buildings and biology students should understand the forces at work in muscles, bones, and joints. These forces create torques, which tell us how the forces affect an objects equilibrium and rate of rotation.

We will find that an object remains in a state of uniform rotational motion unless acted on by a net torque. This principle is the equivalent to Newton's first law. Further the angular acceleration of an object is proportional to the net torque acting on it, which is the analog of Newton's Second Law. A net torque acting on an object causes a change in its rotational energy.

Finally, torques applied to an object through a given time interval can change the objects angular momentum. In the absence of external torques, angular momentum is conserved, a property that explains some of the mysterious and formidable properties of pulsars—remnants of supernova explosions that rotate at equatorial speeds approaching that of light.

Rotational Variables

Angular Position

The Figure shows a reference line, foxed in the body, perpendicular to the rotation axis and rotating with the body. The angular position of this line is the angle of the line relative to a fixed direction, which we take as the zero angular posisition. From geometry, we know that è is given by

$$\theta = \frac{s}{r}$$

Here s is the length of a circular arc that extends from the x axis (the zero angular position) to the reference line, and r is the radius of the circle. An angle defined in this way is measured in radians (rad) rather than in revolutions (Rev) or degrees. The radian, being the ratio of two lengths, is a pure number and thus has no dimension. Because the circumference of a circle or radius r is 2 ð r there are 2 ð radians in a complete cirle:

$$1\,rev = 360^{o} = \frac{2\pi r}{r} = 2\pi\,rad$$

and thus

$$1 \text{ rad} = 57.3^{o} = 0.159 \text{ rev}$$

We do not reset è {theta} to zero with each complete rotation of the reference line about the rotation axis. If the reference line completes two revolutions from the zero angular position, then the angular position è of the line is è = 4 ð r a d

Translation and Rotation

A rigid body is defined as an object in which the relative separations of the component particles are unchanged, the displacement of any one particle with respect to any other particle in the body being fixed. No truly rigid body exists: external forces can deform any solid. For purposed, then, a rigid body is a solid in which the internal forces between the fundamental particles change so drastically with a variation in their relative separation that large forces are required to deform it appreciably.

Up to this point, we have studied the kinematics and dynamics of a particle whose position in three-dimensional space is completely specified by three coordinates. To describe a change in the position of a body of finite extent, such as a rigid body, is much more complicated. For convenience, it is regarded as a combination of two distinct types of motion: translational motion and rotational motion. Purely translational motion occurs if every particle of the body has the same instantaneous velocity as every other particle, the path traced out by some particle being exactly the same as the path traced out by every other particle in the body. Under translational motion, the change in the position of a rigid body is specified completely by three coordinates such as x, y, and z giving the displacement of some one point, such as the center of mass, fixed to the rigid body.

Purely rotational motion occurs if every particle in the body moves in a circle about a single line. This line is called the axis of rotation. Then the radius vectors from the axis to all particles undergo the same angular displacement in the same time. The axis of rotation need not be with the body. In general, any rotation can be specified completely by the three angular displacements with respect to the rectangular-coordinate axes x, y, and z. Any change in the position of the rigid body is thus completely described by three translational and three rotational coordinates.

Any displacement of a rigid body may be arrived at by first displacing the body translationally, without rotation; or conversely, first a rotation and then a displacement. We already know that for any collection of b'—particles—whether at rest with respect to one another, as in a rigid body, or in relative motion, as in the exploding fragments of a shell—the motion of the center of mass is completely determined by the resultant of the external forces acting on the system of particles is

$$\sum F_{net} = Ma_{cm}$$

where M is the total mass of the system and a is the acceleration of the center of mass. There remains the matter of describing rotation of the body about the center of mass and relating it to the external forces acting on the body. We shall find that the kinematics of rotation motion have many similarities the kinematics of translational motion; moreover, the dynamics of rotational motion involve forms of Newton's second law of motion and of the work-energy theorem that are altogether analogous to those used in particle dynamics.

Speed of rotation, angular acceleration, and torque

The speed of rotation is given by the angular frequency (rads) or frequencyrotational speed revolutions per minute (turnss, turnsmin), or period (seconds, days, etc.).With one direction of rotation considered positive, the sign of the angular frequency indicates the direction of rotation. The time-rate of change of angular frequency is the scalar version of angular acceleration (rad/s^2). This change is caused by the scalar version of the torque, which can have a positive or negative value in accordance with the convention of positive and negative angular frequency. The ratio of torque and angular acceleration (how heavy is it to start, stop, or otherwise change rotation) is given by the moment of inertia. The energy required for / released during rotation is the torque times the rotation angle, the energy stored in a rotating object is one half of the moment of inertia times the square of the angular frequency. The power required for angular acceleration is the torque times the angular frequency.

Vectors

According to the right-hand rule, moving away from the observer is associated with clockwise rotation and moving towards the observer with counterclockwise rotation, like most screws. The angular velocityvector also describes the direction of the ax s of rotation. In the case of a fixed axis this direction is along that axis and the rotation process is described by a scalar, the angular frequency, as a function of time.

Similarly the torque vector also describes around which axis it tends to cause rotation, or in other words, the direction in which it tends to change the angular velocity vector. To maintain rotation around the fixed axis the total force has to be zero and the total torque vector has to be along the axis, so that it only changes the magnitude and not the direction of the angular velocity vector. In the case of a hinge, only the component of the torque vector along the axis has effect on the rotation, other forces and torques are compensated by the structure.

Centripetal force

In the case of a spinning object, internal tensile stress provides the centripetal force that keeps the object together. A rigid body model neglects the accompanying strain. If the body is not rigid this strain constitutes a change of shape. This is also expressed as changing shape due to the "centrifugal force". Celestial bodies rotating about each other often have elliptic orbits. The special case of a circular orbits is an example of a rotation around a fixed axis: this axis is the line through the center of mass perpendicular to the plane of motion. The centripetal force is provided by gravity, celestial body, so it need not be solid to keep together, unless the angular speed is too high in relation to its density. For example, a spinning celestial body of water must take at least 3 hours and 18 minutes to rotate, regardless of size, or the water will separate. If the density of the fluid is higher the time can be less. See orbital period.

Constant angular speed

The simplest case of rotation around a fixed axis is that of constant angular speed. The total torque is zero: in e.g. the case of the rotation of the Earth around its axis there is very little friction, in the case of e.g. a fan the motor applies a torque to compensate for friction. The angle of rotation is a linear function of time, which modulo 360° is a periodic function. An example of this is the two-body problem with circular orbits.

Newton's law of universal gravitation

Newton's law of universal gravitation is usually stated that every particle attracts every other particle in the universe with a force which is directly proportional to the product of their masses and inversely proportional to the square of the distance between their centers. This is a general physical law derived from empirical observations by what Isaac Newton called inductive reasoning. It is a part of classical mechanics and was formulated in Newton's work Philosophiæ Naturalis Principia Mathematica ("the Principia"), first published on 5 July 1687. When Newton presented Book 1 of the unpublished text in April 1686 to the Royal Society, Robert Hooke made a claim that Newton had obtained the inverse square law from him. In today's language, the law states that every pointmass attracts every other point mass by a force acting along the line intersecting the two points. The force is proportional to the product of the two masses, and inversely proportional to the square of the distance between them.

The equation for universal gravitation thus takes the form:

$$F = G\frac{m_1 m_2}{r^2},$$

whereF is the gravitational force acting between two objects, m and m are the masses of the objects, r is the distance between the centers of their masses, and G is the gravitational constant. The first test of Newton's theory of gravitation between masses in the laboratory was the Cavendish experiment conducted by the British scientist Henry Cavendish in 1798. It took place 111 years after the publication of Newton's Principia and approximately 71 years after his death.

Newton's law of gravitation resembles Coulomb's law of electrical forces, which is used to calculate the magnitude of the electrical force arising between two charged bodies. Both are inverse-square laws, where force is inversely proportional to the square of the distance between the bodies. Coulomb's law has the product of two charges in place of the product of the masses, and the electrostatic constant in place of the gravitational constant.

Newton's law has since been superseded by Albert Einstein's theory of general relativity, but it continues to be used as an excellent approximation of the effects of gravity in most applications. Relativity is required only when there is a need for extreme accuracy, or when dealing with very strong gravitational fields, such as those found near extremely massive and dense objects, or at very close distances (such as Mercury's orbit around the Sun).

Early history

The relation of the distance of objects in free fall to the square of the time taken had recently been confirmed by Grimaldi and Riccioli between 1640 and 1650. They had also made a calculation of the gravitational constant by recording the oscillations of a pendulum. A modern assessment about the early history of the inverse square law is that "by the late 1670s", the assumption of an "inverse proportion between gravity and the square of distance was rather common and had been advanced by a number of different people for different reasons". The same author credits Robert Hooke with a significant and seminal contribution, but treats Hooke's claim of priority on the inverse square point as irrelevant, as several individuals besides Newton and Hooke had suggested it. He points instead to the idea of "compounding the celestial motions" and the conversion of Newton's thinking away from "centrifugal" and towards "centripetal" force as Hooke's significant contributions.

Newton gave credit in his Principia to two people: Bullialdus (who wrote without proof that there was a force on the Earth towards the Sun), and Borelli (who wrote that all planets were attracted towards the Sun). The main influence may have been Borelli, whose book Newton had a copy of.

Plagiarism dispute

In 1686, when the first book of Newton's Principia was presented to the Royal Society, Robert Hooke accused Newton of plagiarism by claiming that he had taken from him the "notion" of "the rule of the decrease of Gravity, being reciprocally as the squares of the distances from the Center". At the same time (according to Edmond Halley's contemporary report) Hooke agreed that "the Demonstration of the Curves generated thereby" was wholly Newton's. In this way, the question arose as to what, if anything, Newton owed to Hooke. This is a subject extensively discussed since that time and on which some points, outlined below, continue to excite controversy.

Hooke's work and claims

Robert Hooke published his ideas about the "System of the World" in the 1660s, when he read to the Royal Society on March 21, 1666, a paper "concerning the inflection of a direct motion into a curve by a supervening attractive principle", and he published them again in somewhat developed form in 1674, as an addition to "An Attempt to Prove the Motion of the Earth from Observations". Hooke announced in 1674 that he planned to "explain a System of the World differing in many particulars from any yet known", based on three suppositions: that "all Celestial Bodies whatsoever, have an attraction or gravitating power towards their own Centers" and "also attract all the other Celestial Bodies that are within the sphere of their activity"; that "all bodies whatsoever that are put into a direct and simple motion, will so continue to move forward in a straight line, till they are by some other effectual powers deflected and bent..." and that "these attractive powers are so much the more powerful in operating, by how much the nearer the body wrought upon is to their own Centers". Thus Hooke postulated mutual attractions between the Sun and planets, in a way that increased with nearness to the attracting body, together with a principle of linear inertia.

Hooke's statements up to 1674 made no mention, however, that an inverse square law applies or might apply to these attractions. Hooke's gravitation was also not yet universal, though it approached universality more closely than previous hypotheses. He also did not provide accompanying evidence or mathematical demonstration. On the latter two aspects, Hooke himself stated in 1674: "Now what these several degrees [of attraction] are I have not yet experimentally verified"; and as to his whole proposal: "This I only hint at present", "having my self many other things in hand which I would first compleat, and therefore cannot so well attend it" (i.e. "prosecuting this Inquiry"). It was later on, in writing on 6 January 1679|80 to Newton, that Hooke communicated his "supposition... that the Attraction always is in a duplicate proportion to the Distance from the Center Reciprocall, and Consequently that the Velocity will be in a subduplicate proportion to the Attraction and Consequently as Kepler Supposes Reciprocall to the Distance." (The inference about the velocity was incorrect.)

Hooke's correspondence with Newton during 1679-1680 not only mentioned this inverse square supposition for the decline of attraction with increasing distance, but also, in Hooke's opening letter to Newton, of 24 November 1679, an approach of "compounding the celestial motions of the planets of a direct motion by the tangent & an attractive motion towards the central body".

Newton's work and claims

Newton, faced in May 1686 with Hooke's claim on the inverse square law, denied that Hooke was to be credited as author of the idea. Among the reasons, Newton recalled that the idea had been discussed with Sir Christopher Wren previous to Hooke's 1679 letter. Newton also pointed out and acknowledged prior work of others, including Bullialdus, (who suggested, but without demonstration, that there was an attractive force from the Sun in the inverse square proportion to the distance), and Borelli (who suggested, also without demonstration, that there was a centrifugal tendency in counterba lance with a gravitational attraction towards the Sun so as to make the planets move in ellipses). D T Whiteside has described the contribution to Newton's thinking that came from Borelli's book, a copy of which was in Newton's library at his death.

Newton further defended his work by saying that had he first heard of the inverse square proportion from Hooke, he would still have some rights to it in view of his demonstrations of its accuracy. Hooke, without evidence in favor of the supposition, could only guess that the inverse square law was approximately valid at great distances from the center. According to Newton, while the 'Principia' was still at pre-publication stage, there were so many a-priori reasons to doubt the accuracy of the inverse-square law (especially close to an attracting sphere) that "without my (Newton's) Demonstrations, to which Mr Hooke is yet a stranger, it cannot believed by a judicious Philosopher to be any where accurate."

This remark refers among other things to Newton's finding, supported by mathematical demonstration, that if the inverse square law applies to tiny particles, then even a large spherically symmetrical mass also attracts masses external to its surface, even close up, exactly as if all its own mass were concentrated at its center. Thus Newton gave a justification, otherwise lacking, for applying the inverse square law to large spherical planetary masses as if they were tiny particles. In addition, Newton had formulated, in Propositions 43-45 of Book 1 and associated sections of Book 3, a sensitive test of the accuracy of the inverse square law, in which he showed that only where the law of force is calculated as the inverse square of the distance will the directions of orientation of the planets' orbital ellipses stay constant as they are observed to do apart from small effects attributable to inter-planetary perturbations.

In regard to evidence that still survives of the earlier history, manuscripts written by Newton in the 1660s show that Newton himself had, by 1669, arrived at proofs that in a circular case of planetary motion, "endeavour to recede" (what was later called centrifugal force) had an inverse-square relation with distance from the center. After his 1679-1680 correspondence with Hooke, Newton adopted the language of inward or centripetal force. According to Newton scholar J. Bruce Brackenridge, although much has been made of the change in language and difference of point of view, as between centrifugal or centripetal forces, the actual computations and proofs remained the same either way. They also involved the combination of tangential and radial displacements, which Newton was making in the 1660s. The lesson offered by Hooke to Newton here, although significant, was one of perspective and did not change the analysis. This background shows there was basis for Newton to deny deriving the inverse square law from Hooke.

Newton's acknowledgment

On the other hand, Newton did accept and acknowledge, in all editions of the Principia, that Hooke (but not exclusively Hooke) had separately appreciated the inverse square law in the solar system. Newton acknowledged Wren, Hooke and Halley in this connection in the Scholium to Proposition 4 in Book 1. Newton also acknowledged to Halley that his correspondence with Hooke in 1679-80 had reawakened his dormant interest in astronomical matters, but that did not mean, according to Newton, that Hooke had told Newton anything new or original: "yet am I not beholden to him for any light into that business but only for the diversion he gave me from my other studies to think on these things & for his dogmaticalness in writing as if he had found the motion in the Ellipsis, which inclined me to try it..."

Modern priority controversy

Since the time of Newton and Hooke, scholarly discussion has also touched on the question of whether Hooke's 1679 mention of 'compounding the motions' provided Newton with something new and valuable, even though that was not a claim actually voiced by Hooke at the time. As described above, Newton's manuscripts of the 1660s do show him actually combining tangential motion with the effects of radially directed force or endeavour, for example in his derivation of the inverse square relation for the circular case. They also show Newton clearly expressing the concept of linear inertia-for which he was indebted to Descartes' work, published in 1644 (as Hooke probably was). These matters do not appear to have been learned by Newton from Hooke.

Nevertheless, a number of authors have had more to say about what Newton gained from Hooke and some aspects remain controversial. The fact that most of Hooke's private papers had been destroyed or have disappeared does not help to establish the truth. Newton's role in relation to the inverse square law was not as it has sometimes been represented. He did not claim to think it up as a bare idea. What Newton did was to show how the inverse-square law of attraction had many necessary mathematical connections with observable features of the motions of bodies in the solar system; and that they were related in such a way that the observational evidence and the mathematical demonstrations, taken together, gave reason to believe that the inverse square law was not just approximately true but exactly true (to the accuracy achievable in Newton's time and for about two centuries afterwards - and with some loose ends of points that could not yet be certainly examined, where the implications of the theory had not yet been adequately identified or calculated).

About thirty years after Newton's death in 1727, Alexis Clairaut, a mathematical astronomer eminent in his own right in the field of gravitational studies, wrote after reviewing what Hooke published, that "One must not think that this idea... of Hooke diminishes Newton's glory"; and that "the example of Hooke" serves "to show what a distance there is between a truth that is glimpsed and a truth that is demonstrated".

Modern form

In modern language, the law states the following:

Every pointmass attracts every single other point mass by a force acting along the line intersecting both points. The force is proportional to the product of the two masses and inversely proportional to the square of the distance between them:

$$F = G\frac{m_1 m_2}{r^2}$$

where:

- F is the force between the masses;
- G is the gravitational constant (6.674×10 N?·?(m/kg));
- m is the first mass;
- m is the second mass;
- r is the distance between the centers of the masses

Assuming SI units, F is measured in newtons (N), m and m in kilograms (kg), r in meters (m), and the constant G is approximately equal to 6.674×10 N m kg. The value of the constant G was first accurately determined from the results of the Cavendish experiment conducted by the British scientist Henry Cavendish in 1798, although Cavendish did not himself calculate a numerical value for G. This experiment was also the first test of Newton's theory of gravitation between masses in the laboratory. It took place 111 years after the publication of Newton's Principia and 71 years after Newton's death, so none of Newton's calculations could use the value of G; instead he could only calculate a force relative to another force.

Bodies with spatial extent

If the bodies in question have spatial extent (as opposed to being point masses), then the gravitational force between them is calculated by summing the contributions of the notional point masses which constitute the bodies. In the limit, as the component point masses become "infinitely small", this entails integrating the force (in vector form, see below) over the extents of the two bodies.

In this way, it can be shown that an object with a spherically-symmetric distribution of mass exerts the same gravitational attraction on external bodies as if all the object's mass were concentrated at a point at its center. (This is not generally true for non-spherically-symmetrical bodies.) For points inside a spherically-symmetric distribution of matter, Newton's Shell theorem can be used to find the gravitational force. The theorem tells us how different parts of the mass distribution affect the gravitational force measured at a point located a distance r from the center of the mass distribution:

- The portion of the mass that is located at radii r <r causes the same force at r as if all of the mass enclosed within a sphere of radius r was concentrated at the center of the mass distribution (as noted above).
- The portion of the mass that is located at radii r >r exerts no net gravitational force at the distance r from the center. That is, the individual gravitational forces exerted by the elements of the sphere out there, on the point at r, cancel each other out.

As a consequence, for example, within a shell of uniform thickness and density there is no net gravitational acceleration anywhere within the hollow sphere. Furthermore, inside a uniform sphere the gravity increases linearly with the distance from the center; the increase due to the additional mass is 1.5 times the decrease due to the larger distance from the center. Thus, if a spherically symmetric body has a uniform core and a uniform mantle with a density that is less than 2/3 of that of the core, then the gravity initially decreases outwardly beyond the boundary, and if the sphere is large enough, further outward the gravity increases again, and eventually it exceeds the gravity at the core/mantle boundary. The gravity of the Earth may be highest at the core/mantle boundary.

Theoretical concerns with Newton's expression

- There is no immediate prospect of identifying the mediator of gravity. Attempts by physicists to identify the relationship between the gravitational force and other known fundamental forces are not yet resolved, although considerable headway has been made over the last 50 years (See: Theory of everything and Standard Model). Newton himself felt that the concept of an inexplicable action at a distance was unsatisfactory (see "Newton's reservations" below), but that there was nothing more that he could do at the time.
- Newton's theory of gravitation requires that the gravitational force be transmitted instantaneously. Given the classical assumptions of the nature of space and time before the development of General Relativity, a significant propagation delay in gravity leads to unstable planetary and stellar orbits.

Observations conflicting with Newton's formula

- Newton's theory does not fully explain the precession of the perihelion of the orbits of the planets, especially that of Mercury, which was detected long after the life of Newton. There is a 43 arcsecond per century discrepancy between the Newtonian calculation, which arises only from the gravitational attractions from the other planets, and the observed precession, made with advanced telescopes during the 19th century.
- The predicted angular deflection of light rays by gravity that is calculated by using Newton's Theory is only one-half of the deflection that is actually observed by astronomers. Calculations using general relativity are in much closer agreement with the astronomical observations.
- In spiral galaxies, the orbiting of stars around their centers seems to strongly disobey Newton's law of universal gravitation. Astrophysicists, however, explain this spectacular phenomenon in the framework of Newton's laws, with the presence of large amounts of dark matter.

Newton's reservations

While Newton was able to formulate his law of gravity in his monumental work, he was deeply uncomfortable with the notion of "action at a distance" that his equations implied. In 1692, in his third letter to Bentley, he wrote: "That one body may act upon another at a distance through a vacuum without the mediation of anything else, by and through which their action and force may be conveyed from one another, is to me so great an absurdity that, I believe, no man who has in philosophic matters a competent faculty of thinking could ever fall into it."

He never, in his words, "assigned the cause of this power". In all other cases, he used the phenomenon of motion to explain the origin of various forces acting on bodies, but in the case of gravity, he was unable to experimentally identify the motion that produces the force of gravity (although he invented two mechanical hypotheses in 1675 and 1717). Moreover, he refused to even offer a hypothesis as to the cause of this force on grounds that to do so was contrary to sound science. He lamented that "philosophers have hitherto attempted the search of nature in vain" for the source of the gravitational force, as he was convinced "by many reasons" that there were "causes hitherto unknown" that were fundamental to all the "phenomena of nature". These fundamental phenomena are still under investigation and, though hypotheses abound, the definitive answer has yet to be found. And in Newton's 1713 General Scholium in the second edition of Principia: "I have not yet been able to discover the cause of these properties of gravity from phenomena and I feign no hypotheses.... It is enough that gravity does really exist and acts according to the laws I have explained, and that it abundantly serves to account for all the motions of celestial bodies."

Einstein's solutionThese objections were explained by Einstein's theory of general relativity, in which gravitation is an attribute of curved spacetime instead of being due to a force propagated between bodies. In Einstein's theory, energy and momentum distort spacetime in their vicinity, and other particles move in trajectories determined by the geometry of spacetime. This allowed a description of the motions of light and mass that was consistent with all available observations. In general relativity, the gravitational force is a fictitious force due to the curvature of spacetime, because the gravitational acceleration of a body in free fall is due to its world line being a geodesic of spacetime

Solutions of Newton's law of universal gravitation

The n-body problem is an ancient, classical problem of predicting the individual motions of a group of celestial objects interacting with each other gravitationally.

Solving this problem - from the time of the Greeks and on - has been motivated by the desire to understand the motions of the Sun, planets and the visible stars. In the 20th century, understanding the dynamics of globular cluster star systems became an important n-body problem too. The n-body problem in general relativity is considerably more difficult to solve.

The classical physical problem can be informally stated as: given the quasi-steady orbital properties (instantaneous position, velocity and time)of a group of celestial bodies, predict their interactive forces; and consequently, predict their true orbital motions for all future times. The two-body problem has been completely solved, as has the restricted three-body problem.

3

SIMPLE HARMONIC MOTION

In mechanics and physics, simple harmonic motion is a type of periodic motion where the restoring force is directly proportional to the displacement and acts in the direction opposite to that of displacement. Simple harmonic motion can serve as a mathematical model of a variety of motions, such as the oscillation of a spring. In addition, other phenomena can be approximated by simple harmonic motion, including the motion of a simple pendulum as well as molecular vibration. Simple harmonic motion is typified by the motion of a mass on a spring when it is subject to the linear elastic restoring force given by Hooke's Law. The motion is sinusoidal in time and demonstrates a single resonant frequency. In order for simple harmonic motion to take place, the net force of the object at the end of the pendulum must be proportional to the displacement.

Simple harmonic motion provides a basis for the characterization of more complicated motions through the techniques of Fourier analysis.

INTRODUCTION

In the diagram a simple harmonic oscillator, comprising a weight attached to one end of a spring, is shown. The other end of the spring is connected to a rigid support such as a wall. If the system is left at rest at the equilibrium position then there is no net force acting on the mass. However, if the mass is displaced from the equilibrium position, a restoring elastic force which obeys Hooke's law is exerted by the spring.

Mathematically, the restoring force F is given by

where F is the restoring elastic force exerted by the spring (in SI units: N), k is the spring constant (N·m?1), and x is thedisplacement from the equilibrium position (in m).

For any simple harmonic oscillator:

- When the system is displaced from its equilibrium position, a restoring force which resembles Hooke's law tends to restore the system to equilibrium.

Once the mass is displaced from its equilibrium position, it experiences a net restoring force. As a result, it accelerates and starts going back to the equilibrium position. When the mass moves closer to the equilibrium position, the restoring force decreases. At the equilibrium position, the net restoring force vanishes. However, at x = 0, the mass has momentum because of the impulse that the restoring force has imparted. Therefore, the mass continues past the equilibrium position, compressing the spring. A net restoring force then tends to slow it down, until its velocity reaches zero, whereby it will attempt to reach equilibrium position again.

As long as the system has no energy loss, the mass will continue to oscillate. Thus, simple harmonic motion is a type of periodic motion.

Dynamics of simple harmonic motion

For one-dimensional simple harmonic motion, the equation of motion, which is a second-order linear ordinary differential equation with constant coefficients, could be obtained by means of Newton's second law (and Hooke's law for a mass on a spring).

$$F_{net} = m\frac{d^2x}{dt^2} = -kx,$$

where m is the inertial mass of the oscillating body, x is its displacement from the equilibrium (or mean) position, and k is a constant (the spring constant for a mass on a spring).

Therefore, $$\frac{d^2x}{dt^2} = -\left(\frac{k}{m}\right)x,$$

Solving the differential equation above, a solution which is a sinusoidal function is obtained

$$x(t) = c_1 \cos(\omega t) + c_2 \sin(\omega t) = A\cos(\omega t - \varphi),$$

where $\omega = \sqrt{\frac{k}{m}}, A = \sqrt{c_1{}^2 + c_2{}^2}, \tan\varphi = \left(\frac{c_2}{c_1}\right),$

In the solution, c1 and c2 are two constants determined by the initial conditions, and the origin is set to be the equilibrium position. [A] Each of these constants carries a physical meaning of the motion: A is the amplitude (maximum displacement from the equilibrium position), ? = 2?f is the angular frequency, and ? is the phase. [B]

Using the techniques of calculus, the velocity and acceleration as a function of time can be found:

$$v(t) = \frac{\mathrm{d}x}{\mathrm{d}t} = -A\omega \sin(\omega t - \varphi),$$

$$\text{Speed: } \omega\sqrt{A^2 - x^2}$$

Maximum speed (at equilibrium point).

$$a(t) = \frac{\mathrm{d}^2x}{\mathrm{d}t^2} = -A\omega^2 \cos(\omega t - \varphi).$$

Maximum acceleration = Aw^2 (at extreme points)

By definition, if a mass m is under SHM its acceleration is directly proportional to displacement.

$$a(x) = -\omega^2 x. \quad \omega^2 = \left(\frac{k}{m}\right)$$

where $\omega^2 = \left(\frac{k}{m}\right)$

Since ? = 2?f, $f = \frac{1}{2\pi}\sqrt{\frac{k}{m}},$

and, since T = 1/f where T is the time period,

$$T = 2\pi\sqrt{\frac{m}{k}}.$$

These equations demonstrate that the simple harmonic motion is isochronous (the period and frequency are independent of the amplitude and the initial phase of the motion).

Energy

The kinetic energy K of the system at time t is

$$K(t) = \frac{1}{2}mv^2(t) = \frac{1}{2}m\omega^2 A^2 \sin^2(\omega t - \varphi) = \frac{1}{2}kA^2 \sin^2(\omega t - \varphi),$$

and the potential energy is

The total mechanical energy of the system therefore has the constant value

$$E = K + U = \frac{1}{2}kA^2.$$

Examples

The following physical systems are some examples of simple harmonic oscillator.

Mass on a spring

A mass m attached to a spring of spring constant k exhibits simple harmonic motion in closed space. The equation

$$T = 2\pi\sqrt{\frac{m}{k}}$$

shows that the period of oscillation is independent of both the amplitude and gravitational acceleration. The above equation is also valid in the case when a constant force is being applied on the mass, i. e. a constant force cannot change the period of oscillation.

Uniform circular motion

Simple harmonic motion can in some cases be considered to be the one-dimensional projection of uniform circular motion. If an object moves with angular speed ? around a circle of radius r centered at the origin of the x-y plane, then its motion along each coordinate is simple harmonic motion with amplitude r and angular frequency ?.

Mass of a simple pendulum

In the small-angle approximation, the motion of a simple pendulum is approximated by simple harmonic motion. The period of a mass attached to a pendulum of length ? with gravitational acceleration g is given by

$$T = 2\pi\sqrt{\frac{\ell}{g}}$$

This shows that the period of oscillation is independent of the amplitude and mass of the pendulum but not the acceleration due to gravity (g), therefore a pendulum of the same length on the Moon would swing more slowly due to the Moon's lower gravitational field strength.

This approximation is accurate only for small angles because of the expression for angular acceleration ? being proportional to the sine of the displacement angle:

$$mg\ell \sin(\theta) = Ia,$$

where is the moment of inertia. When ? is small, sin?? ? ? and therefore the expression becomes

$$-mg_{\ell}\theta = Ia$$

which makes angular acceleration directly proportional to ?, satisfying the definition of simple harmonic motion.

Scotch yoke

A Scotch yoke mechanism can be used to convert between rotational motion and linear reciprocating motion. The linear motion can take various forms depending on the shape of the slot, but the basic yoke with a constant rotation speed produces a linear motion that is simple harmonic in form.

Applications

This setup is most commonly used in control valve actuators in high-pressure oil and gas pipelines.

Although not a common metalworking machine nowadays, crude shapers can use Scotch yokes. Almost all those use a Whitworth linkage, which gives a slow speed forward cutting stroke and a faster return.

It has been used in various internal combustion engines, such as the Bourke engine, SyTech engine, and many hot air engines and steam engines.

The term scotch yoke continues to be used when the slot in the yoke is shorter than the diameter of the circle made by thecrank pin. For example, the side rods of a locomotive may have scotch yokes to permit vertical motion of intermediate driving axles.

Internal combustion engine uses

Under ideal engineering conditions, force is applied directly in the line of travel of the assembly. The sinusoidal motion, cosinusoidal velocity, and sinusoidal acceleration (assuming constant angular velocity) result in smoother operation. The higher percentage of time spent at top dead centre (dwell) improves theoretical engine efficiency of constant volume combustion cycles. It allows the elimination of joints typically served by a wrist pin, and near elimination of piston skirts and cylinder scuffing, as side loading of piston due to sine of connecting rod angle is mitigated. The longer the distance between the piston and the yoke, the less wear that occurs, but greater the inertia, making such increases in the piston rod length realistically only suitable for lower RPM (but higher torque) applications.

The Scotch Yoke is not used in most internal combustion engines because of the rapid wear of the slot in the yoke caused by sliding friction and high contact pressures. This is mitigated by a sliding block between the crank and the slot in the piston rod. Also, increased heat loss during combustion due to extended dwell at top dead centre offsets any constant volume combustion improvements in real engines. In an engine application, less percent of the time is spent at bottom dead centre when compared to a conventional piston and crankshaft mechanism, which reduces blowdown time for two-stroke engines. Experiments have shown that extended dwell time does not work well with constant volume combustion Otto Cycle Engines. Gains might be more apparent in Otto Cycle Engines using a stratified direct injection (diesel or similar) cycle to reduce heat losses.

Harmonic oscillator

In classical mechanics, a harmonic oscillator is a system that, when displaced from its equilibrium position, experiences a restoring force, F, proportional to the displacement, x: $\vec{F} = -k\vec{x}$

where k is a positive constant.

If F is the only force acting on the system, the system is called a simple harmonic oscillator, and it undergoes simple harmonic motion: sinusoidal oscillations about the equilibrium point, with a constant amplitude and a constant frequency (which does not depend on the amplitude).

If a frictional force (damping) proportional to the velocity is also present, the harmonic oscillator is described as a damped oscillator. Depending on the friction coefficient, the system can:

- Oscillate with a frequency lower than in the non-damped case, and an amplitude decreasing with time (underdamped oscillator).
- Decay to the equilibrium position, without oscillations (overdamped oscillator).

The boundary solution between an underdamped oscillator and an overdamped oscillator occurs at a particular value of the friction coefficient and is called "critically damped. "

If an external time dependent force is present, the harmonic oscillator is described as a driven oscillator.

Mechanical examples include pendulums (with small angles of displacement), masses connected to springs, and acoustical systems. Other analogous systems include electrical harmonic oscillators such as RLC circuits. The harmonic oscillator model is very important in physics, because any mass subject to a force in stable equilibrium acts as a harmonic oscillator for small vibrations. Harmonic oscillators occur widely in nature and are exploited in many manmade devices, such as clocks and radio circuits. They are the source of virtually all sinusoidal vibrations and waves.

Simple harmonic oscillator

A simple harmonic oscillator is an oscillator that is neither driven nor damped. It consists of a mass m, which experiences a single force, F, which pulls the mass in the direction of the point x=0 and depends only on the mass's position x and a constant k. Balance of forces (Newton's second law) for the system is

$$F = ma = m\frac{\mathrm{d}^2x}{\mathrm{d}t^2} = m\ddot{x} = -kx.$$

Solving this differential equation, we find that the motion is described by the function

$$x(t) = A\cos\left(\omega t + \phi\right),$$

where $\omega = \sqrt{\frac{k}{m}} = \frac{2\pi}{T}.$

The motion is periodic, repeating itself in a sinusoidal fashion with constant amplitude, A. In addition to its amplitude, the motion of a simple harmonic oscillator is characterized by its period T, the time for a single oscillation or its frequency f = 1?T, the number of cycles per unit time. The position at a given time t also depends on the phase, ?, which determines the starting point on the sine wave. The period and frequency are determined by the size of the mass m and the force constant k, while the amplitude and phase are determined by the starting position and velocity.

The velocity and acceleration of a simple harmonic oscillator oscillate with the same frequency as the position but with shifted phases. The velocity is maximum for zero displacement, while the acceleration is in the opposite direction as the displacement.

The potential energy stored in a simple harmonic oscillator at position x is

$$U = \frac{1}{2}kx^2.$$

Damped harmonic oscillator

In real oscillators, friction, or damping, slows the motion of the system. Due to frictional force, the velocity decreases in proportion to the acting frictional force.

While simple harmonic motion oscillates with only the restoring force acting on the system, damped harmonic motion experiences friction. In many vibrating systems the frictional force Ff can be modeled as being proportional to the velocity v of the object:Ff = ?cv, where c is called the viscous damping coefficient.

Balance of forces (Newton's second law) for damped harmonic oscillators is then

$$F = F_{ext} - kx - c\frac{\mathrm{d}x}{\mathrm{d}t} = m\frac{\mathrm{d}^2x}{\mathrm{d}t^2}.$$

When no external forces are present (i. e. when), this can be rewritten into the form

$$\frac{\mathrm{d}^2x}{\mathrm{d}t^2} + 2\zeta\omega_0\frac{\mathrm{d}x}{\mathrm{d}t} + \omega_0^2 x = 0,$$

where $\omega_0 = \sqrt{\frac{k}{m}}$

is called the 'undamped angular frequency of the oscillator' and $\zeta = \frac{c}{2\sqrt{mk}}$ is called the 'damping ratio'.

The value of the damping ratio ? critically determines the behavior of the system. A damped harmonic oscillator can be:

- Overdamped The system returns (exponentially decays) to steady state without oscillating. Larger values of the damping ratio ?return to equilibrium slower.
- Critically damped: The system returns to steady state as quickly as possible without oscillating (although overshoot can occur). This is often desired for the damping of systems such as doors.
- Underdamped : The system oscillates (with a slightly different frequency than the undamped case) with the amplitude gradually decreasing to zero. The angular frequency of the underdamped harmonic oscillator is given by

$$\omega_1 = \omega_0\sqrt{1-\zeta^2}.$$

The Q factor of a damped oscillator is defined as

$$Q = 2\pi \times \frac{\text{Energy stored}}{\text{Energy lost per cycle}}.$$

Q is related to the damping ratio by the equation

$$Q = \frac{1}{2\zeta}.$$

Driven harmonic oscillators

Driven harmonic oscillators are damped oscillators further affected by an externally applied force F(t).

Newton's second law takes the form

$$F(t) - kx - c\frac{\mathrm{d}x}{\mathrm{d}t} = m\frac{\mathrm{d}^2x}{\mathrm{d}t^2}.$$

It is usually rewritten into the form

$$\frac{\mathrm{d}^2x}{\mathrm{d}t^2} + 2\zeta\omega_0\frac{\mathrm{d}x}{\mathrm{d}t} + \omega_0^2 x = \frac{F(t)}{m}.$$

This equation can be solved exactly for any driving force, using the solutions z(t) which satisfy the unforced equation:

$$\frac{\mathrm{d}^2z}{\mathrm{d}t^2} + 2\zeta\omega_0\frac{\mathrm{d}z}{\mathrm{d}t} + \omega_0^2 z = 0,$$

and which can be expressed as damped sinusoidal oscillations,

$$z(t) = A\mathrm{e}^{-\zeta\omega_0 t}\ \sin\left(\sqrt{1-\zeta^2}\ \omega_0 t + \phi\right),$$

in the case where ? ? 1. The amplitude A and phase ? determine the behavior needed to match the initial conditions.

Step input

In the case ? < 1 and a unit step input with x(0) = 0:

$$\frac{F(t)}{m} = \begin{cases} \omega_0^2 & t \geq 0 \\ 0 & t < 0 \end{cases}$$

the solution is:

$$x(t) = 1 - \mathrm{e}^{-\zeta\omega_0 t}\frac{\sin\left(\sqrt{1-\zeta^2}\ \omega_0 t + \varphi\right)}{\sin(\varphi)},$$

with phase ? given by

The time an oscillator needs to adapt to changed external conditions is of the order ? = 1/(??0). In physics, the adaptation is called relaxation, and ? is called the relaxation time.

In electrical engineering, a multiple of ? is called the settling time, i. e. the time necessary to ensure the signal is within a fixed departure from final value, typically within 10%. The term overshoot refers to the extent the maximum response exceeds final value, and undershoot refers to the extent the response falls below final value for times following the maximum response.

Sinusoidal driving force

In the case of a sinusoidal driving force:

$$\frac{d^2x}{dt^2} + 2\zeta\omega_0\frac{dx}{dt} + \omega_0^2 x = \frac{1}{m}F_0\sin(\omega t),$$

where F_0 is the driving amplitude and is the driving frequency for a sinusoidal driving mechanism. This type of system appears in AC driven RLC circuits (resistor-inductor-capacitor) and driven spring systems having internal mechanical resistance or external air resistance.

The general solution is a sum of a transient solution that depends on initial conditions, and a steady state that is independent of initial conditions and depends only on the driving amplitude F_0 driving frequency, undamped angular frequency, and the damping ratio ζ

The steady-state solution is proportional to the driving force with an induced phase change of φ

$$x(t) = \frac{F_0}{mZ_m\omega}\sin(\omega t + \phi)$$

where

$$Z_m = \sqrt{(2\omega_0\zeta)^2 + \frac{1}{\omega^2}(\omega_0^2 - \omega^2)^2}$$

is the absolute value of the impedance or linear response function and

$$\phi = \arctan\left(\frac{2\omega\omega_0\zeta}{\omega^2 - \omega_0^2}\right)$$

is the phase of the oscillation relative to the driving force, if the arctan value is taken to be between -180 degrees and 0 (that is, it represents a phase lag, for both positive and negative values of the arctan's argument).

For a parti $\omega_r = \omega_0\sqrt{1 - 2\zeta^2}$ alled the resonance, or resonant frequency . (for a given F_o is maximum. This resonance effect only occurs when $\zeta < 1/\sqrt{2}$

i. e. for significantly underdamped systems. For strongly underdamped systems the value of the amplitude can become quite large near the resonance frequency.

The transient solutions are the same as the unforced $F_o = 0$

damped harmonic oscillator and represent the systems response to other events that occurred previously. The transient solutions typically die out rapidly enough that they can be ignored.

Parametric oscillators

A parametric oscillator is a driven harmonic oscillator in which the drive energy is provided by varying the parameters of the oscillator, such as the damping or restoring force. A familiar example of parametric oscillation is "pumping" on a playground swing. A person on a moving swing can increase the amplitude of the swing's oscillations without any external drive force (pushes) being applied, by changing the moment of inertia of the swing by rocking back and forth ("pumping") or alternately standing and squatting, in rhythm with the swing's oscillations. The varying of the parameters drives the system. Examples of parameters that may be varied are its resonance frequency and damping.

Parametric oscillators are used in many applications. The classical varactor parametric oscillator oscillates when the diode's capacitance is varied periodically. The circuit that varies the diode's capacitance is called the "pump" or "driver". In microwave electronics, waveguide/YAG based parametric oscillators operate in the same fashion. The designer varies a parameter periodically to induce oscillations.

Parametric oscillators have been developed as low-noise amplifiers, especially in the radio and microwave frequency range. Thermal noise is minimal, since a reactance (not a resistance) is varied. Another common use is frequency conversion, e. g., conversion from audio to radio frequencies. For example, the Optical parametric oscillator converts an inputlaser wave into two output waves of lower frequency ( ).

Parametric resonance occurs in a mechanical system when a system is parametrically excited and oscillates at one of its resonant frequencies. Parametric excitation differs from forcing, since the action appears as a time varying modification on a system parameter. This effect is different from regular resonance because it exhibits the instabilityphenomenon.

Universal oscillator equation

The equation $\frac{\mathrm{d}^2q}{\mathrm{d}\tau^2} + 2\zeta\frac{\mathrm{d}q}{\mathrm{d}\tau} + q = 0$

is known as the universal oscillator equation since all second order linear oscillatory systems can be reduced to this form. This is done through nondimensionalization.

The solution to this differential equation contains two parts, the "transient" and the "steady state".

Transient solution

The solution based on solving the ordinary differential equation is for arbitrary constants c1 and c2

$$q_t(\tau) = \begin{cases} e^{-\zeta\tau}\left(c_1 e^{\tau\sqrt{\zeta^2-1}} + c_2 e^{-\tau\sqrt{\zeta^2-1}}\right) & \zeta > 1 \text{ (overdamping)} \\ e^{-\zeta\tau}(c_1 + c_2\tau) = e^{-\tau}(c_1 + c_2\tau) & \zeta = 1 \text{ (critical damping)} \\ e^{-\zeta\tau}\left[c_1\cos\left(\sqrt{1-\zeta^2}\tau\right) + c_2\sin\left(\sqrt{1-\zeta^2}\tau\right)\right] & \zeta < 1 \text{(underdamping)} \end{cases}$$

The transient solution is independent of the forcing function.

Steady-state solution

Apply the "complex variables method" by solving the auxiliary equation below and then finding the real part of its solution:

$$\frac{\mathrm{d}^2q}{\mathrm{d}\tau^2} + 2\zeta\frac{\mathrm{d}q}{\mathrm{d}\tau} + q = \cos(\omega\tau) + \mathrm{i}\sin(\omega\tau) = e^{\mathrm{i}\omega\tau}.$$

Supposing the solution is of the form $q_s(\tau) = Ae^{\mathrm{i}(\omega\tau+\phi)}$.

Its derivatives from zero to 2nd order are

$$q_s = Ae^{\mathrm{i}(\omega\tau+\phi)}, \quad \frac{\mathrm{d}q_s}{\mathrm{d}\tau} = \mathrm{i}\omega Ae^{\mathrm{i}(\omega\tau+\phi)}, \quad \frac{\mathrm{d}^2q_s}{\mathrm{d}\tau^2} = -\omega^2 Ae^{\mathrm{i}(\omega\tau+\phi)}.$$

Substituting these quantities into the differential equation gives

$$-\omega^2 Ae^{\mathrm{i}(\omega\tau+\phi)} + 2\zeta\mathrm{i}\omega Ae^{\mathrm{i}(\omega\tau+\phi)} + Ae^{\mathrm{i}(\omega\tau+\phi)} = (-\omega^2 A + 2\zeta\mathrm{i}\omega A + A)e^{\mathrm{i}(\omega\tau+\phi)} = e^{\mathrm{i}\omega\tau}.$$

Dividing by the exponential term on the left results in

$$-\omega^2 A + 2\zeta\mathrm{i}\omega A + A = e^{-\mathrm{i}\phi} = \cos\phi - \mathrm{i}\sin\phi.$$

Equating the real and imaginary parts results in two independent equations

$$A(1-\omega^2) = \cos\phi \qquad 2\zeta\omega A = -\sin\phi.$$

Compare this result with the theory section on resonance, as well as the "magnitude part" of the RLC circuit. This amplitude function is particularly important in the analysis and understanding of the frequency response of second-order systems.

Phase part

To solve for ö, divide both equations to get

$$\tan\phi = -\frac{2\zeta\omega}{1-\omega^2} = \frac{2\zeta\omega}{\omega^2-1} \Rightarrow \phi \equiv \phi(\zeta,\omega) = \arctan\left(\frac{2\zeta\omega}{\omega^2-1}\right).$$

This phase function is particularly important in the analysis and understanding of the frequency response of second-order systems.

Full solution

Combining the amplitude and phase portions results in the steady-state solution

$$q_s(\tau) = A(\zeta,\omega)\cos(\omega\tau + \phi(\zeta,\omega)) = A\cos(\omega\tau+\phi).$$

The solution of original universal oscillator equation is a superposition (sum) of the transient and steady-state solutions

For a more complete description of how to solve the above equation, see linear ODEs with constant coefficients.

$$q(\tau) = q_t(\tau) + q_s(\tau).$$

Application to a conservative force

The problem of the simple harmonic oscillator occurs frequently in physics, because a mass at equilibrium under the influence of any conservative force, in the limit of small motions, behaves as a simple harmonic oscillator.

A conservative force is one that has a potential energy function. The potential energy function of a harmonic oscillator is:

$$V(x) = \frac{1}{2}kx^2$$

Given an arbitrary potential energy function V(x), one can do a Taylor expansion in terms of x around an energy minimum ($x = x_0$) to model the behavior of small perturbations from equilibrium.

$$V(x) = V(x_0) + (x-x_0)V'(x_0) + \frac{1}{2}(x-x_0)^2V^{(2)}(x_0) + O(x-x_0)^3$$

Because V(x_0) is a minimum, the first derivative evaluated at x_0 must be zero, so the linear term drops out:

$$V(x) = V(x_0) + \frac{1}{2}(x - x_0)^2 V^{(2)}(x_0) + O(x - x_0)^3$$

The constant term V(x_0) is arbitrary and thus may be dropped, and a coordinate transformation allows the form of the simple harmonic oscillator to be retrieved:

$$V(x) \approx \frac{1}{2}x^2 V^{(2)}(0) = \frac{1}{2}kx^2$$

Thus, given an arbitrary potential energy function V(x) with a non-vanishing second derivative, one can use the solution to the simple harmonic oscillator to provide an approximate solution for small perturbations around the equilibrium point.

EXAMPLES

Simple pendulum

Assuming no damping and small amplitudes, the differential equation governing a simple pendulum is

$$\frac{d^2\theta}{dt^2} + \frac{g}{\ell}\theta = 0.$$

The solution to this equation is given by:

$$\theta(t) = \theta_0 \cos\left(\sqrt{\frac{g}{\ell}}t\right) \qquad |\theta_0| \ll 1$$

where is the largest angle attained by the pendulum. The period, the time for one complete oscillation, is given by divided by whatever is multiplying the time in the argument of the cosine ($\sqrt{\frac{g}{\ell}}$here).

$$T_0 = 2\pi\sqrt{\frac{\ell}{g}} \qquad |\theta_0| \ll 1.$$

Pendulum swinging over turntable

Simple harmonic motion can in some cases be considered to be the one-dimensional projection of two-dimensional circular motion. Consider a long pendulum swinging over the turntable of a record player. On the edge of the turntable there is an object. If the object is viewed from the same level as the turntable, a projection of the motion of the object seems to be moving backwards and forwards on a straight line orthogonal to the view direction, sinusoidally like the pendulum.

Spring/mass system

When a spring is stretched or compressed by a mass, the spring develops a restoring force. Hooke's law gives the relationship of the force exerted by the spring when the spring is compressed or stretched a certain length: F(t) = - kx(t)

where F is the force, k is the spring constant, and x is the displacement of the mass with respect to the equilibrium position. The minus sign in the equation indicates that the force exerted by the spring always acts in a direction that is opposite to the displacement (i. e. the force always acts towards the zero position), and so prevents the mass from flying off to infinity.

By using either force balance or an energy method, it can be readily shown that the motion of this system is given by the following differential equation:

$$F(t) = -kx(t) = m\frac{\mathrm{d}^2}{\mathrm{d}t^2}x(t) = ma.$$

... the latter being Newton's second law of motion.

If the initial displacement is A, and there is no initial velocity, the solution of this equation is given by:

$$x(t) = A\cos\left(\sqrt{\frac{k}{m}}t\right).$$

Given an ideal massless spring, is the mass on the end of the spring. If the spring itself has mass, its effective mass must be included in.

Energy variation in the spring–damping system

In terms of energy, all systems have two types of energy, potential energy and kinetic energy. When a spring is stretched or compressed, it stores elastic potential energy, which then is transferred into kinetic energy. The potential energy within a spring is determined by the equation U - $kx^2/2$

When the spring is stretched or compressed, kinetic energy of the mass gets converted into potential energy of the spring. By conservation of energy, assuming the datum is defined at the equilibrium position, when the spring reaches its maximum potential energy, the kinetic energy of the mass is zero. When the spring is released, it tries to return to equilibrium, and all its potential energy converts to kinetic energy of the mass.

LINEAR MOTION

Linear motion (also called rectilinear motion) is a motion along a straight line, and can therefore be described mathematically using only one spatial dimension. The linear motion can be of two types: uniform linear motion with constant velocity or zero acceleration; non uniform linear motion with variable velocity or non-zero acceleration. The motion of a particle (a point-like object) along a line can be described by its position, which varies with (time). An example of linear motion is an athlete running 100m along a straight track.

Linear motion is the most basic of all motion. According to Newton's first law of motion, objects that do not experience any net force will continue to move in a straight line with a constant velocity until they are subjected to a net force. Under everyday circumstances, external forces such as gravity and friction can cause an object to change the direction of its motion, so that its motion cannot be described as linear.

One may compare linear motion to general motion. In general motion, a particle's position and velocity are described by vectors, which have a magnitude and direction. In linear motion, the directions of all the vectors describing the system are equal and constant which means the objects move along the same axis and do not change direction. The analysis of such systems may therefore be simplified by neglecting the direction components of the vectors involved and dealing only with the magnitude.

Neglecting the rotation and other motions of the Earth, an example of linear motion is the ball thrown straight up and falling back straight down.

Displacement

The motion in which all the particles of a body move through the same distance in the same time is called translatory motion. There are two types of translatory motions: rectilinear motion; curvilinear motion. Since linear motion is a motion in a single dimension, the distance traveled by an object in particular direction is the same as displacement. The SI unit of displacement is the metre. If is the initial position of an object and is the final position, then mathematically the displacement is given by:

The equivalent of displacement in rotational motion is the angular displacement measured in radian. The displacement of an object cannot be greater than the distance. Consider a person travelling to work daily. Overall displacement when he returns home is zero, since the person ends up back where he started, but the distance travelled is clearly not zero.

VELOCITY

Velocity is defined as the rate of change of displacement with respect to time. The SI unit of velocity is ms^{-1} or metre per second.

Average velocity

The average velocity is the ratio of total displacement Δx taken over time interval Δt. Mathematically, it is given by:

$$\mathbf{v}_{\mathbf{av}} = \frac{\Delta x}{\Delta t} = \frac{x_2 - x_1}{t_2 - t_1}$$

where:t_1

is the time at which the object was at position $x_{1/2}$

is the time at which the object was at position x_2

Instantaneous velocity

The instantaneous velocity can be found by differentiating the displacement with respect to time.

$$\mathbf{v} = \lim_{\Delta t \to 0} \frac{\Delta x}{\Delta t} = \frac{dx}{dt}$$

Speed

Speed is the absolute value of velocity i. e. speed is always positive. The unit of speed is metre per second. If v is the speed then,

$$v = |\mathbf{v}| = \left|\frac{dx}{dt}\right|$$

The magnitude of the instantaneous velocity is the instantaneous speed.

Acceleration

Acceleration is defined as the rate of change of velocity with respect to time. Acceleration is the second derivative of displacement i. e. acceleration can be found by differentiating position with respect to time twice or differentiating velocity with respect to time once. The SI unit of acceleration is ms^{-2} or metre per second squared.

If a_{av} is the average acceleration and Δv - v_2 - v_1 is the average velocity over the time interval Δt then mathematically,

$$\mathbf{a_{av}} = \frac{\Delta \mathbf{v}}{\Delta t} = \frac{\mathbf{v_2} - \mathbf{v_1}}{t_2 - t_1}$$

The instantaneous acceleration is the limit of the ratio Δv and Δt as approaches zero i. e.,

$$\mathbf{a} = \lim_{\Delta t \to 0} \frac{\Delta \mathbf{v}}{\Delta t} = \frac{d\mathbf{v}}{dt} = \frac{d^2 x}{dt^2}$$

Jerk

The rate of change of acceleration, the third derivative of displacement is known as jerk. The SI unit of jerk is ms^{-3}. In the UK jerk is also called as jolt.

Jounce

The rate of change of jerk, the fourth derivative of displacement is known as jounce. The SI unit of jounce is ms^{-4} which can be pronounced as metres per quartic second.

Equations of kinematics

In case of constant acceleration, the four physical quantities acceleration, velocity, time and displacement can be related by using the Equations of motionhere, u

is the initial velocity v

is the final velocity a

is the acceleration s

is the displacement t

is the time

These relationships can be demonstrated graphically. The gradient of a line on a displacement time graph represents the velocity. The gradient of the velocity time graph gives the acceleration while the area under the velocity time graph gives the displacement. The area under an acceleration time graph gives the change in velocity.

Analogy between linear and rotational motion

The following table refers to rotation of a rigid body about a fixed axis:s is arclength r, is the distance from the axis to any point, and at is the tangential acceleration, which is the component of the acceleration that is parallel to the motion. In contrast, the centripetal acceleration, $\mathbf{a_c} = v^2/r = \omega^2 r$ is perpendicular to the motion. The component of the force parallel to the motion, or equivalently, perpendicular to the line connecting the point of application to the axis is $F_{\perp}$. The sum is over j = 1 to N particles and/or points of application.

4

WORK, ENERGY, AND POWER

Work is the product of force and displacement. In physics, a force is said to do work if, when acting, there is a movement of the point of application in the direction of the force. For example, when a ball is held above the ground and then dropped, the work done on the ball as it falls is equal to the weight of the ball (a force) multiplied by the distance to the ground (a displacement). When the force F {\displaystyle F is constant and the angle between the force and the displacement sdisplaystyle s} is è, then the work done is given by W = Fs cos è. Work transfers energy from one place to another, or one form to another. The SI unit of work is the joule (J).

ETYMOLOGY

According to Jammer, the term work was introduced in 1826 by the French mathematician Gaspard-Gustave Coriolis as "weight lifted through a height", which is based on the use of early steam engines to lift buckets of water out of flooded ore mines. According to Rene Dugas, French engineer and historian, it is to Solomon of Caux "that we owe the term work in the sense that it is used in mechanics now".

Units

The SI unit of work is the joule (J), which is defined as the work expended by a force of one newton through a displacement of one metre.

The dimensionally equivalent newton-metre (NÅ"m) is sometimes used as the measuring unit for work, but this can be confused with the unit newton-metre, which is the measurement unit of torque. Usage of NÅ"m is discouraged by the SI authority, since it can lead to confusion as to whether the quantity expressed in newton metres is a torque measurement, or a measurement of work. Non-SI units of work include the newton-metre, erg, the foot-pound, the foot-poundal, the kilowatt hour, the litre-atmosphere, and the horsepower-hour. Due to work having the same physical dimension as heat, occasionally measurement units typically reserved for heat or energy content, such as therm, BTU and Calorie, are utilized as a measuring unit.

Work and energy

The work W done by a constant force of magnitude F on a point that moves a displacement s in a straight line in the direction of the force is the product

$$W = F s$$

For example, if a force of 10 newtons (F = 10 N) acts along a point that travels 2 metres (s = 2 m), then W = F s = (10 N) (2 m) = 20 J. This is approximately the work done lifting a 1 kg object from ground level to over a person's head against the force of gravity. The work is doubled either by lifting twice the weight the same distance or by lifting the same weight twice the distance. Work is closely related to energy. The work-energy principle states that an increase in the kinetic energy of a rigid body is caused by an equal amount of positive work done on the body by the resultant force acting on that body. Conversely, a decrease in kinetic energy is caused by an equal amount of negative work done by the resultant force. From Newton's second law, it can be shown that work on a free (no fields), rigid (no internal degrees of freedom) body, is equal to the change in kinetic energy K E of the linear velocity and angular velocity of that body,

$$W = K E.$$

he work of forces generated by a potential function is known as potential energy and the forces are said to be conservative. Therefore, work on an object that is merely displaced in a conservative force field, without change in velocity or rotation, is equal to minus the change of potential energy P E of the object,

$$W = \text{“} P E.$$

These formulas show that work is the energy associated with the action of a force, so work subsequently possesses the physical dimensions, and units, of energy. The work/energy principles discussed here are identical to Electric work/energy principles.

Constraint forces

Constraint forces limit the movement of components in a system, such as constraining an object to a surface (in the case of a slope plus gravity, the object is stuck to the slope, when attached to a taut string it cannot move in an outwards direction to make the string any 'tauter'). Constraint forces restrict the velocity in the direction of the constraint to zero, which means the constraint forces do not perform work on the system.

For a mechanical system, constraint forces eliminate movement in directions that characterize the constraint. Thus constraint forces do not perform work on the system, because the component of velocity along the constraint force at each point of application is zero. For example, in a pulley system like the Atwood machine, the internal forces on the rope and at the supporting pulley do no work on the system. Therefore work need only be computed for the gravity forces acting on the bodies.

For example, the centripetal force exerted inwards by a string on a ball in uniform circular motionsideways constrains the ball to circular motion restricting its movement away from the center of the circle. This force does zero work because it is perpendicular to the velocity of the ball.

Another example is a book on a table. If external forces are applied to the book so that it slides on the table, then the force exerted by the table constrains the book from moving downwards. The force exerted by the table supports the book and is perpendicular to its movement which means that this constraint force does not perform work. The magnetic force on a charged particle is F = qv × B, where q is the charge, v is the velocity of the particle, and B is the magnetic field. The result of a cross product is always perpendicular to both of the original vectors, so F¥"v. The dot product of two perpendicular vectors is always zero, so the work W = FÅ"v = 0, and the magnetic force does not do work. It can change the direction of motion but never change the speed.

Mathematical calculation

For moving objects, the quantity of work/time (power) is integrated along the trajectory of the point of application of the force. Thus, at any instant, the rate of the work done by a force (measured in joules/second, or watts) is the scalar product of the force (a vector), and the velocity vector of the point of application. This scalar product of force and velocity is known as instantaneous power. Just as velocities may be integrated over time to obtain a total distance, by the fundamental theorem of calculus, the total work along a path is similarly the time-integral of instantaneous power applied along the trajectory of the point of application. Work is the result of a force on a point that follows a curve X, with a velocity v, at each instant. The small amount of work äW that occurs over an instant of time dt is calculated as

$$\delta W = F.\ ds = F.\ vdt$$

where the FÅ"v is the power over the instant dt. The sum of these small amounts of work over the trajectory of the point yields the work,

$$W = \int_{t_1}^{t_2} \mathbf{F} \cdot \mathbf{v} dt = \int_{t_1}^{t_2} \mathbf{F} \cdot \frac{d\mathbf{s}}{dt} dt = \int_C \mathbf{F} \cdot d\mathbf{s},$$

whereC is the trajectory from x(t) to x(t). This integral is computed along the trajectory of the particle, and is therefore said to be path dependent.

If the force is always directed along this line, and the magnitude of the force is F, then this integral simplifies to

$$W = \int_C F\,ds$$

wheres is displacement along the line. If F is constant, in addition to being directed along the line, then the integral simplifies further to

$$W = \int_C F\,ds = F\int_C ds = Fs$$

wheres is the displacement of the point along the line.

This calculation can be generalized for a constant force that is not directed along the line, followed by the particle. In this case the dot productFÅ"ds = F cos èds, where è is the angle between the force vector and the direction of movement, that is

$$W = \int_C \mathbf{F}\cdot d\mathbf{s} = Fs\cos\theta.$$

In the notable case of a force applied to a body always at an angle of 90° from the velocity vector (as when a body moves in a circle under a central force), no work is done at all, since the cosine of 90 degrees is zero. Thus, no work can be performed by gravity on a planet with a circular orbit (this is ideal, as all orbits are slightly elliptical). Also, no work is done on a body moving circularly at a constant speed while constrained by mechanical force, such as moving at constant speed in a frictionless ideal centrifuge.

Work done by a variable force

Calculating the work as "force times straight path segment" would only apply in the most simple of circumstances, as noted above. If force is changing, or if the body is moving along a curved path, possibly rotating and not necessarily rigid, then only the path of the application point of the force is relevant for the work done, and only the component of the force parallel to the application point velocity is doing work (positive work when in the same direction, and negative when in the opposite direction of the velocity). This component of force can be described by the scalar quantity called scalar tangential component (F cos a è, where è is the angle between the force and the velocity). And then the most general definition of work can be formulated as follows:

Work of a force is the line integral of its scalar tangential component along the path of its application point. If the force varies (e.g. compressing a spring) we need to use calculus to find the work done. If the force is given by F(x) (a function of x) then the work done by the force along the x-axis from a to b is:

$$W = \int_a^b \mathbf{F}(\mathbf{s})\cdot d\mathbf{s}$$

Torque and rotation

A force couple results from equal and opposite forces, acting on two different points of a rigid body. The sum (resultant) of these forces may cancel, but their effect on the body is the couple or torque T. The work of the torque is calculated as

$$dW = \mathbf{T} \cdot \vec{\omega} dt,$$

where the TÅ"ù is the power over the instant ät. The sum of these small amounts of work over the trajectory of the rigid body yields the work,

$$W = \int_{t_1}^{t_2} \mathbf{T} \cdot \vec{\omega} dt.$$

This integral is computed along the trajectory of the rigid body with an angular velocity ù that varies with time, and is therefore said to be path dependent.

If the angular velocity vector maintains a constant direction, then it takes the form,

$$\vec{\omega} = \dot{\phi} \mathbf{S},$$

where ö is the angle of rotation about the constant unit vector S. In this case, the work of the torque becomes,

$$W = \int_{t_1}^{t_2} \mathbf{T} \cdot \vec{\omega} dt = \int_{t_1}^{t_2} \mathbf{T} \cdot \mathbf{S} \frac{d\phi}{dt} dt = \int_{C} \mathbf{T} \cdot \mathbf{S} d\phi,$$

whereC is the trajectory from ö(t) to ö(t). This integral depends on the rotational trajectory ö(t), and is therefore path-dependent.

If the torque T is aligned with the angular velocity vector so that,

$$T = \tau S,$$

and both the torque and angular velocity are constant, then the work takes the form,This result can be understood more simply by considering the torque as arising from a force of constant magnitude F, being applied perpendicularly to a lever arm at a distance r, as shown in the figure. This force will act through the distance along the circular arc s = rö, so the work done is

$$W = Fs = Fr\varphi.$$

Introduce the torque ô = Fr, to obtain

$$W = Fr\varphi = \tau\varphi,$$

as presented above.

Notice that only the component of torque in the direction of the angular velocity vector contributes to the work.

Work and potential energy

The scalar product of a force F and the velocity v of its point of application defines the power input to a system at an instant of time. Integration of this power over the trajectory of the point of application, C = x(t), defines the work input to the system by the force.

Path dependence

Therefore, the work done by a force F on an object that travels along a curve C is given by the line integral:

$$W = \int_C \mathbf{F} \cdot d\mathbf{x} = \int_{t_1}^{t_2} \mathbf{F} \cdot \mathbf{v}dt,$$

wheredx(t) defines the trajectory C and v is the velocity along this trajectory. In general this integral requires the path along which the velocity is defined, so the evaluation of work is said to be path dependent.

The time derivative of the integral for work yields the instantaneous power,

$$\frac{dW}{dt} = P(t) = \mathbf{F} \cdot \mathbf{v}.$$

Path independence

If the work for an applied force is independent of the path, then the work done by the force, by the gradient theorem, defines a potential function which is evaluated at the start and end of the trajectory of the point of application. This means that there is a potential function U(x), that can be evaluated at the two points x(t) and x(t) to obtain the work over any trajectory between these two points. It is tradition to define this function with a negative sign so that positive work is a reduction in the potential, that is

$$W = \int_C \mathbf{F} \cdot \mathrm{d}\mathbf{x} = \int_{\mathbf{x}(t_1)}^{\mathbf{x}(t_2)} \mathbf{F} \cdot \mathrm{d}\mathbf{x} = U(\mathbf{x}(t_1)) - U(\mathbf{x}(t_2)).$$

The function U(x) is called the potential energy associated with the applied force. The force derived from such a potential function is said to be conservative. Examples of forces that have potential energies are gravity and spring forces.

In this case, the gradient of work yields

$$\nabla W = -\nabla U = -(\frac{\partial U}{\partial x}, \frac{\partial U}{\partial y}, \frac{\partial U}{\partial z}) = \mathbf{F},$$

and the force F is said to be "derivable from a potential."

Because the potential U defines a force F at every point x in space, the set of forces is called a force field. The power applied to a body by a force field is obtained from the gradient of the work, or potential, in the direction of the velocity V of the body, that is

$$P(t) = - \nabla U. v = F.v.$$

Work by gravity

Gravity F = mg does work W = mgh along any descending path In the absence of other forces, gravity results in a constant downward acceleration of every freely moving object. Near Earth's surface the acceleration due to gravity is g = 9.8 mÅ"s and the gravitational force on an object of mass m is F = mg. It is convenient to imagine this gravitational force concentrated at the center of mass of the object. If an object is displaced upwards or downwards a vertical distance y " y, the work W done on the object by its weight mg is:

$$W = F_g (y_2 - y_1) = F_g \Delta_y = -mg\Delta y$$

whereF is weight (pounds in imperial units, and newtons in SI units), and Äy is the change in height y. Notice that the work done by gravity depends only on the vertical movement of the object. The presence of friction does not affect the work done on the object by its weight.

Work by gravity in space

The force of gravity exerted by a mass M on another mass m is given by

$$\mathbf{F} = -\frac{GMm}{r^3}\mathbf{r},$$

wherer is the position vector from M to m.

Let the mass m move at the velocity v then the work of gravity on this mass as it moves from position r(t) to r(t) is given by

$$W = -\int_{\mathbf{r}(t_1)}^{\mathbf{r}(t_2)} \frac{GMm}{r^3}\mathbf{r}\cdot d\mathbf{r} = -\int_{t_1}^{t_2} \frac{GMm}{r^3}\mathbf{r}\cdot\mathbf{v}dt.$$

Notice that the position and velocity of the mass m are given by

$$\mathbf{r} = r\mathbf{e}_r, \qquad \mathbf{v} = \dot{r}\mathbf{e}_r + r\dot{\theta}\mathbf{e}_t,$$

where e and e are the radial and tangential unit vectors directed relative to the vector from M to m. Use this to simplify the formula for work of gravity to,

$$W = -\int_{t_1}^{t_2} \frac{GmM}{r^3}(r\mathbf{e}_r)\cdot(\dot{r}\mathbf{e}_r + r\dot{\theta}\mathbf{e}_t)dt = -\int_{t_1}^{t_2} \frac{GmM}{r^3}r\dot{r}dt = \frac{GMm}{r(t_2)} - \frac{GMm}{r(t_1)}.$$

This calculation uses the fact that

$$\frac{d}{dt}r^{-1} = -r^{-2}\dot{r} = -\frac{\dot{r}}{r^2}.$$

The function

$$U = -\frac{GMm}{r},$$

is the gravitational potential function, also known as gravitational potential energy. The negative sign follows the convention that work is gained from a loss of potential energy.

Work by a spring

Forces in springs assembled in parallel

Consider a spring that exerts a horizontal force F = (“kx, 0, 0) that is proportional to its deflection in the x direction independent of how a body moves. The work of this spring on a body moving along the space with the curve X(t) = (x(t), y(t), z(t)), is calculated using its velocity, v = (v, v, v), to obtain

$$W = \int_0^t \mathbf{F} \cdot \mathbf{v} dt = -\int_0^t kxv_x dt = -\frac{1}{2}kx^2.$$

For convenience, consider contact with the spring occurs at t = 0, then the integral of the product of the distance x and the x-velocity, xv, is (1/2)x. The velocity is not a factor here. The work is the product of the distance times the spring force, which is also dependent on distance; hence the x result.

Work by a gas

$$W = \int_a^b P dV$$

Where P is pressure, V is volume, and a and b are initial and final volumes.

Work–energy principle

The principle of work and kinetic energy (also known as the work–energy principle) states that the work done by all forces acting on a particle (the work of the resultant force) equals the change in the kinetic energy of the particle. That is, the work W done by the resultant force on a particle equals the change in the particle's kinetic energy E k

$$W = \Delta E_k = \tfrac{1}{2}mv_2^2 - \tfrac{1}{2}mv_1^2,$$

wherev displaystyle and displaystyle are the speeds of the particle before and after the work is done, and m is its mass. The derivation of the work–energy principle begins with Newton's second law of motion and the resultant force on a particle. Computation of the scalar product of the forces with the velocity of the particle evaluates the instantaneous power added to the system.

Constraints define the direction of movement of the particle by ensuring there is no component of velocity in the direction of the constraint force. This also means the constraint forces do not add to the instantaneous power. The time integral of this scalar equation yields work from the instantaneous power, and kinetic energy from the scalar product of velocity and acceleration. The fact that the work–energy principle eliminates the constraint forces underlies Lagrangian mechanics.

This section focuses on the work–energy principle as it applies to particle dynamics. In more general systems work can change the potential energy of a mechanical device, the thermal energy in a thermal system, or the electrical energy in an electrical device. Work transfers energy from one place to another or one form to another.

Derivation for a particle moving along a straight line

In the case the resultant forceF is constant in both magnitude and direction, and parallel to the velocity of the particle, the particle is moving with constant acceleration a along a straight line. The relation between the net force and the acceleration is given by the equation F = ma (Newton's second law), and the particle displacements can be expressed by the equation

$$s = \frac{v_2^2 - v_1^2}{2a}$$

which follows from $v_2^2 = v_1^2 + 2as$

The work of the net force is calculated as the product of its magnitude and the particle displacement. Substituting the above equations, one obtains:

$$W = Fs = mas = ma(\frac{v_2^2 - v_1^2}{2a}) = \frac{mv_2^2}{2} - \frac{mv_1^2}{2} = \Delta E_k$$

Other derivation:

$$W = Fs = mas = m(\frac{v_2^2 - v_1^2}{2s})s$$

$$W = -\frac{1}{2}mv^2$$

Vertical displacement derivation

$$W = F \times S = mg \times h$$

In the general case of rectilinear motion, when the net force F is not constant in magnitude, but is constant in direction, and parallel to the velocity of the particle, the work must be integrated along the path of the particle:

$$W = \int_{t_1}^{t_2} \mathbf{F} \cdot \mathbf{v} dt = \int_{t_1}^{t_2} F v dt = \int_{t_1}^{t_2} m a v dt = m \int_{t_1}^{t_2} v \frac{dv}{dt} dt = m \int_{v_1}^{v_2} v dv = \frac{1}{2} m (v_2^2 - v_1^2).$$

Power (physics)

In physics, power is the rate of doing work or of transferring heat, i.e. the amount of energy transferred or converted per unit time. Having no direction, it is a scalar quantity. In the International System of Units, the unit of power is the joule per second (J/s), known as the watt (W) in honour of James Watt, the eighteenth-century developer of the condenser steam engine. Another common and traditional measure is horsepower (comparing to the power of a horse); one mechanical horsepower equals about 745.7 watts. Being the rate of work, the equation for power can be written as:

$$\text{power} = \frac{\text{work}}{\text{time}}$$

As a physical concept, power requires both a change in the physical system and a specified time in which the change occurs. This is distinct from the concept of work, which is measured only in terms of a net change in the state of the physical system. The same amount of work is done when carrying a load up a flight of stairs whether the person carrying it walks or runs, but more power is needed for running because the work is done in a shorter amount of time.

The output power of an electric motor is the product of the torque that the motor generates and the angular velocity of its output shaft. The power involved in moving a ground vehicle is the product of the traction force on the wheels and the velocity of the vehicle. The power of a jet-propelled vehicle is the product of the engine thrust and the velocity of the vehicle. The rate at which a light bulb converts electrical energy into light and heat is measured in watts—the higher the wattage, the more power, or equivalently the more electrical energy is used per unit time.

Units

The dimension of power is energy divided by time. The SI unit of power is the watt (W), which is equal to one joule per second. Other units of power include ergs per second (erg/s), horsepower (hp), metric horsepower (Pferdestärke (PS) or cheval vapeur (CV)), and foot-pounds per minute. Other units include dBm, a logarithmic measure relative to a reference of 1 milliwatt, calories per hour, BTU per hour (BTU/h), and tons of refrigeration.

Equations for power

Power, as a function of time, is the rate (i.e. derivative) at which work is done, so can be expressed by this equation:

P = d W d t

$$P = \frac{dW}{dt}$$

whereP is power, W is work, and t is time. Because work is a forceF applied over a distancex,

W = F.x

for a constant force, power can be rewritten as:

$$P = \frac{dW}{dt} = \frac{d}{dt}(\mathbf{F} \cdot \mathbf{x}) = \mathbf{F} \cdot \frac{d\mathbf{x}}{dt} = \mathbf{F} \cdot \mathbf{v}$$

In fact, this is valid for any force, as a consequence of applying the fundamental theorem of calculus.

Average power

As a simple example, burning one kilogram of coal releases much more energy than does detonating a kilogram of TNT, but because the TNT reaction releases energy much more quickly, it delivers far more power than the coal. If ÄW is the amount of work performed during a period of time of duration Ät, the average powerP over that period is given by the formula:

$$P_{\text{avg}} = \frac{\Delta W}{\Delta t}$$

It is the average amount of work done or energy converted per unit of time. The average power is often simply called "power" when the context makes it clear.

The instantaneous power is then the limiting value of the average power as the time interval Ät approaches zero.

$$P = \lim_{\Delta t \to 0} P_{\text{avg}} = \lim_{\Delta t \to 0} \frac{\Delta W}{\Delta t} = \frac{\mathrm{d}W}{\mathrm{d}t}$$

In the case of constant power P, the amount of work performed during a period of duration t is given by:

W = Pt

In the context of energy conversion, it is more customary to use the symbol E rather than W.

Mechanical power

Power in mechanical systems is the combination of forces and movement. In particular, power is the product of a force on an object and the object's velocity, or the product of a torque on a shaft and the shaft's angular velocity.

Mechanical power is also described as the time derivative of work. In mechanics, the work done by a force F on an object that travels along a curve C is given by the line integral:

$$W_C = \int_C \mathbf{F} \cdot \mathbf{v} dt = \int_C \mathbf{F} \cdot d\mathbf{x}$$

wherex defines the path C and v is the velocity along this path.

If the force F is derivable from a potential (conservative), then applying the gradient theorem (and remembering that force is the negative of the gradient of the potential energy) yields:

$$W_c = U(B)-U(A)$$

whereA and B are the beginning and end of the path along which the work was done.

The power at any point along the curve C is the time derivative:

$$P(t) = \frac{dW}{dt} = \mathbf{F} \cdot \mathbf{v} = -\frac{dU}{dt}$$

In one dimension, this can be simplified to:

$$P(t) = F.v$$

In rotational systems, power is the product of the torqueô and angular velocityù,

$$P(t) = ô Å$$

where ù measured in radians per second. The Å"represents scalar product.

In fluid power systems such as hydraulic actuators, power is given by

$$P(t) = pQ$$

wherep is pressure in pascals, or N/m and Q is volumetric flow rate in m/s in SI units.

Mechanical advantage

If a mechanical system has no losses, then the input power must equal the output power. This provides a simple formula for the mechanical advantage of the system.

Let the input power to a device be a force F acting on a point that moves with velocity v and the output power be a force F acts on a point that moves with velocity v. If there are no losses in the system, then

$$P = F_B v_B = F_A v_A$$

and the mechanical advantage of the system (output force per input force) is given by

$$\mathrm{MA} = \frac{F_B}{F_A} = \frac{v_A}{v_B}$$

The similar relationship is obtained for rotating systems, where T and ù are the torque and angular velocity of the input and T and ù are the torque and angular velocity of the output. If there are no losses in the system, then

$$P = T_A w_A = T_B w_B$$

which yields the mechanical advantage

$$\mathrm{MA} = \frac{T_B}{T_A} = \frac{\omega_A}{\omega_B}$$

These relations are important because they define the maximum performance of a device in terms of velocity ratios determined by its physical dimensions. See for example gear ratios.

Energy

In physics, energy is the quantitativeproperty that must be transferred to an object in order to perform work on, or to heat, the object. Energy is a conserved quantity; the law of conservation of energy states that energy can be converted in form, but not created or destroyed. The SI unit of energy is the joule, which is the energy transferred to an object by the work of moving it a distance of 1 metre against a force of 1 newton.

Common forms of energy include the kinetic energy of a moving object, the potential energy stored by an object's position in a force field (gravitational, electric or magnetic), the elastic energy stored by stretching solid objects, the chemical energy released when a fuel burns, the radiant energy carried by light, and the thermal energy due to an object's temperature.

Mass and energy are closely related. Due to mass–energy equivalence, any object that has mass when stationary (called rest mass) also has an equivalent amount of energy whose form is called rest energy, and any additional energy (of any form) acquired by the object above that rest energy will increase the object's total mass just as it increases its total energy. For example, after heating an object, its increase in energy could be measured as a small increase in mass, with a sensitive enough scale.

Living organisms require energy to stay alive, such as the energy humans get from food. Human civilization requires energy to function, which it gets from energy resources such as fossil fuels, nuclear fuel, or renewable energy. The processes of Earth's climate and ecosystem are driven by the radiant energy Earth receives from the sun and the geothermal energy contained within the earth.

Units of energy

Because energy is defined via work, the SIunit of energy is the same as the unit of work – the joule (J), named in honor of James Prescott Joule and his experiments on the mechanical equivalent of heat. In slightly more fundamental terms, 1 joule is equal to 1 newton metre and, in terms of SI base units

$$1\,\mathrm{J} = 1\,\mathrm{kg}\left(\frac{\mathrm{m}}{\mathrm{s}}\right)^2 = 1\frac{\mathrm{kg}\cdot\mathrm{m}^2}{\mathrm{s}^2}$$

An energy unit that is used in atomic physics, particle physics and high energy physics is the electronvolt (eV). One eV is equivalent to 1.60217653×10 J. In spectroscopy the unit cm = 0.000123986 eV is used to represent energy since energy is inversely proportional to wavelength from the equation E = h í = h c / ë In discussions of energy production and consumption, the units barrel of oil equivalent and ton of oil equivalent are often used. Cubic mile of oil is sometimes used as a unit of energy in discussions of global scale energy economics.

When discussing amounts of energy released in explosions or bolideimpact events, the TNT equivalent unit is often used. The joule is the most used unit of energy.

British Imperial / US Customary units

The British Imperial units and U.S. Customary units for both energy and work include the foot-pound force (1.3558 J), the British thermal unit (BTU) which has various values in the region of 1055 J, the horsepower-hour (2.6845 MJ), and the gasoline gallon equivalent (about 120 MJ).

Electricity

The energy unit used for everyday electricity, particularly for utility bills, is the kilowatt-hour (kWh); one kWh is equivalent to 3.6×10 J (3600 kJ or 3.6 MJ). Electricity usage is often given in units of kilowatt-hours per year (kWh/yr). This is actually a measurement of average power consumption, i.e., the average rate at which energy is transferred. One kWh/yr is about 0.11 watts.

Natural gas

Natural gas in the US is sold in Therms or 100 cubic feet (100 ft = 1 Ccf). One Therm is equal to about 105.5 megajoules. In Australia, natural gas is sold in Cubic Meters (1m = 38 megajoules). In the most of the world, natural gas is sold in gigajoules.

Food industry

The calorie is defined as the amount of thermal energy necessary to raise the temperature of one gram of water by 1 Celsius degree, from a temperature of 14.5

°C, at a pressure of 1 atm. For thermochemistry a calorie of 4.184 J is used, but other calories have also been defined, such as the International Steam Table calorie of 4.1868 J. Food energy is measured in large calories or kilocalories, often simply written capitalized as Calories (= 1000 calories).

Atom physics and chemistry

In physics and chemistry, it is still common to measure energy on the atomic scale in the non-SI, but convenient, unitselectronvolts (eV). The Hartree (the atomic unit of energy) is commonly used in calculations. Historically Rydberg units have been used.

Spectroscopy

In spectroscopy and related fields it is common to measure energy levels in units of reciprocal centimetres. These units (cm"1) are strictly speaking not energy units but units proportional to energies, with being the proportionality constant.

Explosions

A gram of TNT releases 980–1100calories upon explosion. To define the tonne of TNT, this was arbitrarily standardized by letting 1000 thermochemical calories = 1 gram TNT = 4184 J (exactly).

Conservation of energy

In physics and chemistry, the law of conservation of energy states that the total energy of an isolated system remains constant; it is said to be conserved over time. This law means that energy can neither be created nor destroyed; rather, it can only be transformed or transferred from one form to another. For instance, chemical energy is converted to kinetic energy when a stick of dynamite explodes. If one adds up all forms of energy that were released in the explosion, such as the kinetic energy and potential energy of the pieces, as well as heat and sound, one will get the exact decrease of chemical energy in the combustion of the dynamite.

Classically, conservation of energy was distinct from conservation of mass; however, special relativity showed that mass is related to energy and vice versa by E = mc, and science now takes the view that mass–energy as a whole is conserved. Theoretically, this implies that an object with mass can itself be converted to pure energy, and vice versa, though this is believed to be possible only under the most extreme of physical conditions, such as likely existed in the universe very shortly after the big bang.

Conservation of energy can be rigorously proven by Noether's theorem as a consequence of continuoustime translation symmetry; that is, from the fact that the laws of physics do not change over time. A consequence of the law of conservation of energy is that a perpetual motion machine of the first kind cannot exist, that is to say, no system without an external energy supply can deliver an unlimited amount of energy to its surroundings. For systems which do not have time translation symmetry, it may not be possible to define conservation of energy. Examples include curved spacetimes in general relativity or time crystals in condensed matter physics.

Mechanical equivalent of heat

A key stage in the development of the modern conservation principle was the demonstration of the mechanical equivalent of heat. The caloric theory maintained that heat could neither be created nor destroyed, whereas conservation of energy entails the contrary principle that heat and mechanical work are interchangeable. In the middle of the eighteenth century, Mikhail Lomonosov, a Russian scientist, postulated his corpusculo-kinetic theory of heat, which rejected the idea of a caloric. Through the results of empirical studies, Lomonosov came to the conclusion that heat was not transferred through the particles of the caloric fluid.

In 1798, Count Rumford (Benjamin Thompson) performed measurements of the frictional heat generated in boring cannons, and developed the idea that heat is a form of kinetic energy; his measurements refuted caloric theory, but were imprecise enough to leave room for doubt. The mechanical equivalence principle was first stated in its modern form by the German surgeon Julius Robert von Mayer in 1842. Mayer reached his conclusion on a voyage to the Dutch East Indies, where he found that his patients' blood was a deeper red because they were consuming less oxygen, and therefore less energy, to maintain their body temperature in the hotter climate. He discovered that heat and mechanical work were both forms of energy and in 1845, after improving his knowledge of physics, he published a monograph that stated a quantitative relationship between them.

Meanwhile, in 1843, James Prescott Joule independently discovered the mechanical equivalent in a series of experiments. In the most famous, now called the "Joule apparatus", a descending weight attached to a string caused a paddle immersed in water to rotate. He showed that the gravitational potential energy lost by the weight in descending was equal to the internal energy gained by the water through friction with the paddle. Over the period 1840–1843, similar work was carried out by engineer Ludwig A. Colding, although it was little known outside his native Denmark.

Both Joule's and Mayer's work suffered from resistance and neglect but it was Joule's that eventually drew the wider recognition. For the dispute between Joule and Mayer over priority, see Mechanical equivalent of heat: Priority. In 1844, William Robert Grove postulated a relationship between mechanics, heat, light, electricity and magnetism by treating them all as manifestations of a single "force" (energy in modern terms). In 1846, Grove published his theories in his book The Correlation of Physical Forces. In 1847, drawing on the earlier work of Joule, Sadi

Carnot and Émile Clapeyron, Hermann von Helmholtz arrived at conclusions similar to Grove's and published his theories in his book Über die Erhaltung der Kraft (On the Conservation of Force, 1847). The general modern acceptance of the principle stems from this publication.

In 1850, William Rankine first used the phrase the law of the conservation of energy for the principle. In 1877, Peter Guthrie Tait claimed that the principle originated with Sir Isaac Newton, based on a creative reading of propositions 40 and 41 of the Philosophiae Naturalis Principia Mathematica. This is now regarded as an example of Whig history.

MASS–ENERGY EQUIVALENCE

Matter is composed of atoms and what makes up atoms. It has intrinsic or rest mass. In the limited range of recognized experience of the nineteenth century it was found that such rest mass is conserved. Einstein's 1905 theory of special relativity showed that it corresponds to an equivalent amount of rest energy. This means that it can be converted to or from equivalent amounts of other (non-material) forms of energy, for example kinetic energy, potential energy, and electromagnetic radiant energy. When this happens, as recognized in twentieth century experience, rest mass is not conserved, unlike the total mass or total energy. All forms of energy contribute to the total mass and total energy. For example, an electron and a positron each have rest mass. They can perish together, converting their combined rest energy into photons having electromagnetic radiant energy, but no rest mass. If this occurs within an isolated system that does not release the photons or their energy into the external surroundings, then neither the total mass nor the total energy of the system will change. The produced electromagnetic radiant energy contributes just as much to the inertia (and to any weight) of the system as did the rest mass of the electron and positron before their demise. Likewise, non-material forms of energy can perish into matter, which has rest mass.

Thus, conservation of energy (total, including material or rest energy), and conservation of mass (total, not just rest), each still holds as an (equivalent) law. In the 18th century these had appeared as two seemingly-distinct laws.

Conservation of energy in beta decay

The discovery in 1911 that electrons emitted in beta decay have a continuous rather than a discrete spectrum appeared to contradict conservation of energy, under the then-current assumption that beta decay is the simple emission of an electron from a nucleus. This problem was eventually resolved in 1933 by Enrico Fermi who proposed the correct description of beta-decay as the emission of both an electron and an antineutrino, which carries away the apparently missing energy.

Relativity

With the discovery of special relativity by Henri Poincaré and Albert Einstein, energy was proposed to be one component of an energy-momentum 4-vector. Each of the four components (one of energy and three of momentum) of this vector is separately conserved across time, in any closed system, as seen from any given inertial reference frame. Also conserved is the vector length (Minkowski norm), which is the rest mass for single particles, and the invariant mass for systems of particles (where momenta and energy are separately summed before the length is calculated—see the article on invariant mass). The relativistic energy of a single massive particle contains a term related to its rest mass in addition to its kinetic energy of motion. In the limit of zero kinetic energy (or equivalently in the rest frame) of a massive particle, or else in the center of momentum frame for objects or systems which retain kinetic energy, the total energy of particle or object (including internal kinetic energy in systems) is related to its rest mass or its invariant mass via the famous equation E = m c 2 displaystyle E= mc 2.

Thus, the rule of conservation of energy over time in special relativity continues to hold, so long as the reference frame of the observer is unchanged. This applies to the total energy of systems, although different observers disagree as to the energy value. Also conserved, and invariant to all observers, is the invariant mass, which is the minimal system mass and energy that can be seen by any observer, and which is defined by the energy–momentum relation.

In general relativity, energy–momentum conservation is not well-defined except in certain special cases. Energy-momentum is typically expressed with the aid of a stress–energy–momentum pseudotensor. However, since pseudotensors are not tensors, they do not transform cleanly between reference frames. If the metric under consideration is static (that is, does not change with time) or asymptotically flat (that is, at an infinite distance away spacetime looks empty), then energy conservation holds without major pitfalls. In practice, some metrics such as the Friedmann–Lemaître–Robertson–Walker metric do not satisfy these constraints and energy conservation is not well defined. The theory of general relativity leaves open the question of whether there is a conservation of energy for the entire universe.

Quantum theory

In quantum mechanics, energy of a quantum system is described by a self-adjoint (or Hermitian) operator called the Hamiltonian, which acts on the Hilbert space (or a space of wave functions) of the system. If the Hamiltonian is a time-independent operator, emergence probability of the measurement result does not change in time over the evolution of the system. Thus the expectation value of energy is also time independent. The local energy conservation in quantum field theory is ensured by the quantum Noether's theorem for energy-momentum tensor operator.

Due to the lack of the (universal) time operator in quantum theory, the uncertainty relations for time and energy are not fundamental in contrast to the position-momentum uncertainty principle, and merely holds in specific cases (see Uncertainty principle). Energy at each fixed time can in principle be exactly measured without any trade-off in precision forced by the time-energy uncertainty relations. Thus the conservation of energy in time is a well defined concept even in quantum mechanics.

Potential energy

In physics, potential energy is the energy held by an object because of its position relative to other objects, stresses within itself, its electric charge, or other factors. Common types of potential energy include the gravitational potential energy of an object that depends on its mass and its distance from the center of mass of another object, the elastic potential energy of an extended spring, and the electric potential energy of an electric charge in an electric field. The unit for energy in the International System of Units (SI) is the joule, which has the symbol J.

The term potential energy was introduced by the 19th-century Scottish engineer and physicist William Rankine, although it has links to Greek philosopher Aristotle's concept of potentiality. Potential energy is associated with forces that act on a body in a way that the total work done by these forces on the body depends only on the initial and final positions of the body in space. These forces, that are called conservative forces, can be represented at every point in space by vectors expressed as gradients of a certain scalar function called potential. Since the work of potential forces acting on a body that moves from a start to an end position is determined only by these two positions, and does not depend on the trajectory of the body, there is a function known as potential that can be evaluated at the two positions to determine this work.

Overview

There are various types of potential energy, each associated with a particular type of force. For example, the work of an elastic force is called elastic potential energy; work of the gravitational force is called gravitational potential energy; work of the Coulomb force is called electric potential energy; work of the strong nuclear force or weak nuclear force acting on the baryoncharge is called nuclear potential energy; work of intermolecular forces is called intermolecular potential energy. Chemical potential energy, such as the energy stored in fossil fuels, is the work of the Coulomb force during rearrangement of mutual positions of electrons and nuclei in atoms and molecules. Thermal energy usually has two components: the kinetic energy of random motions of particles and the potential energy of their mutual positions. Forces derivable from a potential are also called conservative forces. The work done by a conservative force is

$$W = - \Delta U$$

where Ä U is the change in the potential energy associated with the force. The negative sign provides the convention that work done against a force field increases potential energy, while work done by the force field decreases potential energy. Common notations for potential energy are PE, U, V, and E.

Potential energy is the energy by virtue of an object's position relative to other objects. Potential energy is often associated with restoring forces such as a spring or the force of gravity. The action of stretching a spring or lifting a mass is performed by an external force that works against the force field of the potential. This work is stored in the force field, which is said to be stored as potential energy. If the external force is removed the force field acts on the body to perform the work as it moves the body back to the initial position, reducing the stretch of the spring or causing a body to fall. Consider a ball whose mass is m and whose height is h. The acceleration g of free fall is approximately constant, so the weight force of the ball mg is constant. Force × displacement gives the work done, which is equal to the gravitational potential energy, thus U g = m g h

Work and potential energy

Potential energy is closely linked with forces. If the work done by a force on a body that moves from A to B does not depend on the path between these points (if the work is done by a conservative force), then the work of this force measured from A assigns a scalar value to every other point in space and defines a scalar potential field. In this case, the force can be defined as the negative of the vector gradient of the potential field.

If the work for an applied force is independent of the path, then the work done by the force is evaluated at the start and end of the trajectory of the point of application. This means that there is a function U(x), called a "potential," that can be evaluated at the two points x and x to obtain the work over any trajectory between these two points. It is tradition to define this function with a negative sign so that positive work is a reduction in the potential, that is

$$W = \int_C \mathbf{F} \cdot d\mathbf{x} = U(\mathbf{x}_A) - U(\mathbf{x}_B)$$

Gravitational potential energy

Gravitational energy is the potential energy associated with gravitational force, as work is required to elevate objects against Earth's gravity. The potential energy due to elevated positions is called gravitational potential energy, and is evidenced by water in an elevated reservoir or kept behind a dam. If an object falls from one point to another point inside a gravitational field, the force of gravity will do positive work on the object, and the gravitational potential energy will decrease by the same amount.

Consider a book placed on top of a table. As the book is raised from the floor to the table, some external force works against the gravitational force. If the book falls back to the floor, the "falling" energy the book receives is provided by the gravitational force. Thus, if the book falls off the table, this potential energy goes to accelerate the mass of the book and is converted into kinetic energy. When the book hits the floor this kinetic energy is converted into heat, deformation, and sound by the impact.

The factors that affect an object's gravitational potential energy are its height relative to some reference point, its mass, and the strength of the gravitational field it is in. Thus, a book lying on a table has less gravitational potential energy than the same book on top of a taller cupboard and less gravitational potential energy than a heavier book lying on the same table. An object at a certain height above the Moon's surface has less gravitational potential energy than at the same height above the Earth's surface because the Moon's gravity is weaker. "Height" in the common sense of the term cannot be used for gravitational potential energy calculations when gravity is not assumed to be a constant. The following sections provide more detail.

General formula

However, over large variations in distance, the approximation that g is constant is no longer valid, and we have to use calculus and the general mathematical definition of work to determine gravitational potential energy. For the computation of the potential energy, we can integrate the gravitational force, whose magnitude is given by Newton's law of gravitation, with respect to the distance r between the two bodies. Using that definition, the gravitational potential energy of a system of masses m and M at a distance r using gravitational constantG is

$$U = -G\frac{m_1 M_2}{r} + K$$

whereK is an arbitrary constant dependent on the choice of datum from which potential is measured. Choosing the convention that K=0 (i.e. in relation to a point at infinity) makes calculations simpler, albeit at the cost of making U negative; for why this is physically reasonable, see below. Given this formula for U, the total potential energy of a system of n bodies is found by summing, for all n (n " 1) 2 pairs of two bodies, the potential energy of the system of those two bodies.

Considering the system of bodies as the combined set of small particles the bodies consist of, and applying the previous on the particle level we get the negative gravitational binding energy. This potential energy is more strongly negative than the total potential energy of the system of bodies as such since it also includes the negative gravitational binding energy of each body. The potential energy of the system of bodies as such is the negative of the energy needed to separate the bodies from each other to infinity, while the gravitational binding energy is the energy needed to separate all particles from each other to infinity.

U = " m (G M 1 r 1 + G M 2 r 2)

$$U = -m(G\frac{M_1}{r_1} + G\frac{M_2}{r_2})$$

therefore,

$$U = -m\sum G\frac{M}{r}$$

Negative gravitational energy

As with all potential energies, only differences in gravitational potential energy matter for most physical purposes, and the choice of zero point is arbitrary. Given that there is no reasonable criterion for preferring one particular finite r over another, there seem to be only two reasonable choices for the distance at which U becomes zero. The choice of at infinity may seem peculiar, and the consequence that gravitational energy is always negative may seem counterintuitive, but this choice allows gravitational potential energy values to be finite, albeit negative. The singularity at r = 0 {\displaystyle r=0} in the formula for gravitational potential energy means that the only other apparently reasonable alternative choice of convention, with U = 0 {\displaystyle U=0} for r = 0 {\displaystyle r=0}, would result in potential energy being positive, but infinitely large for all nonzero values of r, and would make calculations involving sums or differences of potential energies beyond what is possible with the real number system. Since physicists abhor infinities in their calculations, and r is always non-zero in practice, the choice of U = 0 {\displaystyle U=0} at infinity is by far the more preferable choice, even if the idea of negative energy in a gravity well appears to be peculiar at first. The negative value for gravitational energy also has deeper implications that make it seem more reasonable in cosmological calculations where the total energy of the universe can meaningfully be considered; seeinflation theory for more on this.

Uses

Gravitational potential energy has a number of practical uses, notably the generation of pumped-storage hydroelectricity. For example, in Dinorwig, Wales, there are two lakes, one at a higher elevation than the other. At times when surplus electricity is not required (and so is comparatively cheap), water is pumped up to the higher lake, thus converting the electrical energy (running the pump) to gravitational potential energy. At times of peak demand for electricity, the water flows back down through electrical generator turbines, converting the potential energy into kinetic energy and then back into electricity. The process is not completely efficient and some of the original energy from the surplus electricity is in fact lost to friction.

Gravitational potential energy is also used to power clocks in which falling weights operate the mechanism. It's also used by counterweights for lifting up an elevator, crane, or sash window. Roller coasters are an entertaining way to utilize potential energy - chains are used to move a car up an incline (building up gravitational potential energy), to then have that energy converted into kinetic energy as it falls. Another practical use is utilizing gravitational potential energy to descend (perhaps coast) downhill in transportation such as the descent of an automobile, truck, railroad train, bicycle, airplane, or fluid in a pipeline. In some cases the kinetic energy obtained from the potential energy of descent may be used to start ascending the next grade such as what happens when a road is undulating and has frequent dips. The commercialization of stored energy (in the form of rail cars raised to higher elevations) that is then converted to electrical energy when needed by an electrical grid, is being undertaken in the United States in a system called Advanced Rail Energy Storage (ARES).

Chemical potential energy

Chemical potential energy is a form of potential energy related to the structural arrangement of atoms or molecules. This arrangement may be the result of chemical bonds within a molecule or otherwise. Chemical energy of a chemical substance can be transformed to other forms of energy by a chemical reaction. As an example, when a fuel is burned the chemical energy is converted to heat, same is the case with digestion of food metabolized in a biological organism. Green plants transform solar energy to chemical energy through the process known as photosynthesis, and electrical energy can be converted to chemical energy through electrochemical reactions. The similar term chemical potential is used to indicate the potential of a substance to undergo a change of configuration, be it in the form of a chemical reaction, spatial transport, particle exchange with a reservoir, etc.

Electric potential energy

An object can have potential energy by virtue of its electric charge and several forces related to their presence. There are two main types of this kind of potential energy: electrostatic potential energy, electrodynamic potential energy (also sometimes called magnetic potential energy).

Electrostatic potential energy

Electrostatic potential energy between two bodies in space is obtained from the force exerted by a charge Q on another charge q which is given by

$$\mathbf{F}_e = -\frac{1}{4\pi\varepsilon_0}\frac{Qq}{r^2}\hat{\mathbf{r}},$$

where r ^ is a vector of length 1 pointing from Q to q and å is the vacuum permittivity. This may also be written using Coulomb's constantk = 1 D 4.

If the electric charge of an object can be assumed to be at rest, then it has potential energy due to its position relative to other charged objects. The electrostatic potential energy is the energy of an electrically charged particle (at rest) in an electric field. It is defined as the work that must be done to move it from an infinite distance away to its present location, adjusted for non-electrical forces on the object. This energy will generally be non-zero if there is another electrically charged object nearby.

The work W required to move q from A to any point B in the electrostatic force field is given by

$$\Delta U_{AB}(\mathbf{r}) = -\int_A^B \mathbf{F}_\mathrm{e} \cdot d\mathbf{r}$$

typically given in J for Joules. A related quantity called electric potential (commonly denoted with a V for voltage) is equal to the electric potential energy per unit charge.

Magnetic potential energy

The energy of a magnetic moment? in an externally produced magnetic B-fieldB has potential energyThe magnetizationM in a field is

$$U = -\frac{1}{2}\int \mathbf{M}\cdot\mathbf{B}\,\mathrm{d}V,$$

where the integral can be over all space or, equivalently, where M is nonzero. Magnetic potential energy is the form of energy related not only to the distance between magnetic materials, but also to the orientation, or alignment, of those materials within the field. For example, the needle of a compass has the lowest magnetic potential energy when it is aligned with the north and south poles of the Earth's magnetic field. If the needle is moved by an outside force, torque is exerted on the magnetic dipole of the needle by the Earth's magnetic field, causing it to move back into alignment. The magnetic potential energy of the needle is highest when its field is in the same direction as the Earth's magnetic field. Two magnets will have potential energy in relation to each other and the distance between them, but this also depends on their orientation. If the opposite poles are held apart, the potential energy will be higher the further they are apart and lower the closer they are. Conversely, like poles will have the highest potential energy when forced together, and the lowest when they spring apart.Nuclear potential energy

Nuclear potential energy is the potential energy of the particles inside an atomic nucleus. The nuclear particles are bound together by the strong nuclear force. Weak nuclear forces provide the potential energy for certain kinds of radioactive decay, such as beta decay.

Nuclear particles like protons and neutrons are not destroyed in fission and fusion processes, but collections of them can have less mass than if they were individually free, in which case this mass difference can be liberated as heat and radiation in nuclear reactions (the heat and radiation have the missing mass, but it often escapes from the system, where it is not measured). The energy from the Sun is an example of this form of energy conversion. In the Sun, the process of hydrogen fusion converts about 4 million tonnes of solar matter per second into electromagnetic energy, which is radiated into space.

Kinetic energy

]In physics, the kinetic energy (KE) of an object is the energy that it possesses due to its motion. It is defined as the work needed to accelerate a body of a given mass from rest to its stated velocity. Having gained this energy during its acceleration, the body maintains this kinetic energy unless its speed changes. The same amount of work is done by the body when decelerating from its current speed to a state of rest.

In classical mechanics, the kinetic energy of a non-rotating object of massm traveling at a speedv is In relativistic mechanics, this is a good approximation only when v is much less than the speed of light. The standard unit of kinetic energy is the joule, while the imperial unit of kinetic energy is the foot-pound.

Overview

Energy occurs in many forms, including chemical energy, thermal energy, electromagnetic radiation, gravitational energy, electric energy, elastic energy, nuclear energy, and rest energy. These can be categorized in two main classes: potential energy and kinetic energy. Kinetic energy is the movement energy of an object. Kinetic energy can be transferred between objects and transformed into other kinds of energy.

Kinetic energy may be best understood by examples that demonstrate how it is transformed to and from other forms of energy. For example, a cyclist uses chemical energy provided by food to accelerate a bicycle to a chosen speed. On a level surface, this speed can be maintained without further work, except to overcome air resistance and friction. The chemical energy has been converted into kinetic energy, the energy of motion, but the process is not completely efficient and produces heat within the cyclist.

The kinetic energy in the moving cyclist and the bicycle can be converted to other forms. For example, the cyclist could encounter a hill just high enough to coast up, so that the bicycle comes to a complete halt at the top. The kinetic energy has now largely been converted to gravitational potential energy that can be released by freewheeling down the other side of the hill. Since the bicycle lost some of its energy to friction, it never regains all of its speed without additional pedaling. The energy is not destroyed; it has only been converted to another form by friction. Alternatively, the cyclist could connect a dynamo to one of the wheels and generate some electrical energy on the descent. The bicycle would be traveling slower at the bottom of the hill than without the generator because some of the energy has been diverted into electrical energy. Another possibility would be for the cyclist to apply the brakes, in which case the kinetic energy would be dissipated through friction as heat.

Like any physical quantity that is a function of velocity, the kinetic energy of an object depends on the relationship between the object and the observer's frame of reference. Thus, the kinetic energy of an object is not invariant.

Spacecraft use chemical energy to launch and gain considerable kinetic energy to reach orbital velocity. In an entirely circular orbit, this kinetic energy remains constant because there is almost no friction in near-earth space. However, it becomes apparent at re-entry when some of the kinetic energy is converted to heat. If the orbit is elliptical or hyperbolic, then throughout the orbit kinetic and potential energy are exchanged; kinetic energy is greatest and potential energy lowest at closest approach to the earth or other massive body, while potential energy is greatest and kinetic energy the lowest at maximum distance. Without loss or gain, however, the sum of the kinetic and potential energy remains constant.

Kinetic energy can be passed from one object to another. In the game of billiards, the player imposes kinetic energy on the cue ball by striking it with the cue stick. If the cue ball collides with another ball, it slows down dramatically, and the ball it hit accelerates its speed as the kinetic energy is passed on to it. Collisions in billiards are effectively elastic collisions, in which kinetic energy is preserved. In inelastic collisions, kinetic energy is dissipated in various forms of energy, such as heat, sound, binding energy (breaking bound structures).

Flywheels have been developed as a method of energy storage. This illustrates that kinetic energy is also stored in rotational motion.

Several mathematical descriptions of kinetic energy exist that describe it in the appropriate physical situation. For objects and processes in common human experience, the formula $\frac{1}{2}mv^2$ given by Newtonian (classical) mechanics is suitable. However, if the speed of the object is comparable to the speed of light, relativistic effects become significant and the relativistic formula is used. If the object is on the atomic or sub-atomic scale, quantum mechanical effects are significant, and a quantum mechanical model must be employed.

Kinetic energy of systems

A system of bodies may have internal kinetic energy due to the relative motion of the bodies in the system. For example, in the Solar System the planets and planetoids are orbiting the Sun. In a tank of gas, the molecules are moving in all directions. The kinetic energy of the system is the sum of the kinetic energies of the bodies it contains.

A macroscopic body that is stationary (i.e. a reference frame has been chosen to correspond to the body's center of momentum) may have various kinds of internal energy at the molecular or atomic level, which may be regarded as kinetic energy, due to molecular translation, rotation, and vibration, electron translation and spin, and nuclear spin. These all contribute to the body's mass, as provided by the special theory of relativity. When discussing movements of a macroscopic body, the kinetic energy referred to is usually that of the macroscopic movement only. However all internal energies of all types contribute to body's mass, inertia, and total energy.

Mechanical energy

In physical sciences, mechanical energy is the sum of potential energy and kinetic energy. It is the energy associated with the motion and position of an object. The principle of conservation of mechanical energy states that in an isolated system that is only subject to conservative forces, the mechanical energy is constant. If an object moves in the opposite direction of a conservative net force, the potential energy will increase; and if the speed (not the velocity) of the object changes, the kinetic energy of the object also changes. In all real systems, however, nonconservative forces, such as frictional forces, will be present, but if they are of negligible magnitude, the mechanical energy changes little and its conservation is a useful approximation. In elastic collisions, the mechanical energy is conserved, but in inelastic collisions some mechanical energy is converted into thermal energy. The equivalence between lost mechanical energy (dissipation) and an increase in temperature was discovered by James Prescott Joule. Many devices are used to convert mechanical energy to or from other forms of energy, e.g. an electric motor converts electrical energy to mechanical energy, an electric generator converts mechanical energy into electrical energy and a heat engine converts heat energy to mechanical energy.

General

Energy is a scalar quantity and the mechanical energy of a system is the sum of the potential energy (which is measured by the position of the parts of the system) and the kinetic energy (which is also called the energy of motion):

$$E_{mechanical} = U + K$$

The potential energy, U, depends on the position of an object subjected to a conservative force. It is defined as the object's ability to do work and is increased as the object is moved in the opposite direction of the direction of the force. If F represents the conservative force and x the position, the potential energy of the force between the two positions x and x is defined as the negative integral of F from x to x:The kinetic energy, K, depends on the speed of an object and is the ability of a moving object to do work on other objects when it collides with them. It is defined as one half the product of the object's mass with the square of its speed, and the total kinetic energy of a system of objects is the sum of the kinetic energies of the respective objects:The principle of conservation of mechanical energy states that if a body or system is subjected only to conservative forces, the mechanical energy of that body or system remains constant. The difference between a conservative and a non-conservative force is that when a conservative force moves an object from one point to another, the work done by the conservative force is independent of the path. On the contrary, when a non-conservative force acts upon an object, the work done by the non-conservative force is dependent of the path.Conservation of mechanical energy

According to the principle of conservation of mechanical energy, the mechanical energy of an isolated system remains constant in time, as long as the system is free of friction and other non-conservative forces. In any real situation, frictional forces and other non-conservative forces are present, but in many cases their effects on the system are so small that the principle of conservation of mechanical energy can be used as a fair approximation. Though energy cannot be created or destroyed in an isolated system, it can be converted to another form of energy.Swinging pendulum

In a mechanical system like a swinging pendulum subjected to the conservative gravitational force where frictional forces like air drag and friction at the pivot are negligible, energy passes back and forth between kinetic and potential energy but never leaves the system. The pendulum reaches greatest kinetic energy and least potential energy when in the vertical position, because it will have the greatest speed and be nearest the Earth at this point. On the other hand, it will have its least kinetic energy and greatest potential energy at the extreme positions of its swing, because it has zero speed and is farthest from Earth at these points. However, when taking the frictional forces into account, the system loses mechanical energy with each swing because of the negative work done on the pendulum by these non-conservative forces.Conversion

Today, many technological devices convert mechanical energy into other forms of energy or vice versa. These devices can be placed in these categories:

- An electric motor converts electrical energy into mechanical energy.
- A generator converts mechanical energy into electrical energy.

- A hydroelectric powerplant converts the mechanical energy of water in a storage dam into electrical energy.
- An internal combustion engine is a heat engine that obtains mechanical energy from chemical energy by burning fuel. From this mechanical energy, the internal combustion engine often generates electricity.
- A steam engine converts the heat energy of steam into mechanical energy.
- A turbine converts the kinetic energy of a stream of gas or liquid into mechanical energy.

5

LINEAR MOMENTUM

In Newtonian mechanics, linear momentum, translational momentum, or simply momentum (pl. momenta) is the product of the mass and velocity of an object. It is a vector quantity, possessing a magnitude and a direction. If m is an object's mass and v is its velocity (also a vector quantity), then the object's momentum is:

$$P = mv.$$

In SI units, momentum is measured in kilogram meters per second (kgÅ"m/s).

Newton's second law of motion states that a body's rate of change in momentum is equal to the net force acting on it. Momentum depends on the frame of reference, but in any inertial frame it is a conserved quantity, meaning that if a closed system is not affected by external forces, its total linear momentum does not change. Momentum is also conserved in special relativity (with a modified formula) and, in a modified form, in electrodynamics, quantum mechanics, quantum field theory, and general relativity. It is an expression of one of the fundamental symmetries of space and time: translational symmetry.

Advanced formulations of classical mechanics, Lagrangian and Hamiltonian mechanics, allow one to choose coordinate systems that incorporate symmetries and constraints. In these systems the conserved quantity is generalized momentum, and in general this is different from the kinetic momentum defined above. The concept of generalized momentum is carried over into quantum mechanics, where it becomes an operator on a wave function. The momentum and position operators are related by the Heisenberg uncertainty principle.

In continuous systems such as electromagnetic fields, fluids and deformable bodies, a momentum density can be defined, and a continuum version of the conservation of momentum leads to equations such as the Navier–Stokes equations for fluids or the Cauchy momentum equation for deformable solids or fluids.

NEWTONIAN

Momentum is a vector quantity: it has both magnitude and direction. Since momentum has a direction, it can be used to predict the resulting direction and speed of motion of objects after they collide. Below, the basic properties of momentum are described in one dimension. The vector equations are almost identical to the scalar equations (see multiple dimensions).

Single particle

The momentum of a particle is conventionally represented by the letter p. It is the product of two quantities, the particle's mass (represented by the letter m) and its velocity (v):

$$p = mv.$$

The unit of momentum is the product of the units of mass and velocity. In SI units, if the mass is in kilograms and the velocity is in meters per second then the momentum is in kilogram meters per second (kgÅ"m/s). In cgs units, if the mass is in grams and the velocity in centimeters per second, then the momentum is in gram centimeters per second (gÅ"cm/s).

Being a vector, momentum has magnitude and direction. For example, a 1 kg model airplane, traveling due north at 1 m/s in straight and level flight, has a momentum of 1 kgÅ"m/s due north measured with reference to the ground.

Many particles

The momentum of a system of particles is the vector sum of their momenta. If two particles have respective masses m and m, and velocities v and v, the total momentum is

$$\begin{aligned} p &= p_1 + p_2 \\ &= m_1 v_1 + m_2 v_2 . \end{aligned}$$

The momenta of more than two particles can be added more generally with the following:

$$p = \sum_i m_i v_i .$$

A system of particles has a center of mass, a point determined by the weighted sum of their positions:

If one or more of the particles is moving, the center of mass of the system will generally be moving as well (unless the system is in pure rotation around it). If the total mass of the particles is m and the center of mass is moving at velocity v, the momentum of the system is:

$$p = mv_{cm}.$$

This is known as Euler's first law.

Relation to force

If the net force F applied to a particle is constant, and is applied for a time interval Ät, the momentum of the particle changes by an amount

$$\Delta p = F\Delta t.$$

In differential form, this is Newton's second law; the rate of change of the momentum of a particle is equal to the instantaneous force F acting on it,

$$F = \frac{dp}{dt}.$$

If the net force experienced by a particle changes as a function of time, F(t), the change in momentum (or impulse J) between times t and t is

$$\Delta p = J = \int_{t_1}^{t_2} F(t)dt.$$

Impulse is measured in the derived units of the newton second (1 NÅs = 1 kgÅm/s) or dyne second (1 dyneÅs = 1 gÅcm/s) Under the assumption of constant mass m, it is equivalent to write

$$F = \frac{d(mv)}{dt} = m\frac{dv}{dt} = ma,$$

hence the net force is equal to the mass of the particle times its acceleration.

Example: A model airplane of mass 1 kg accelerates from rest to a velocity of 6 m/s due north in 2 s. The net force required to produce this acceleration is 3 newtons due north. The change in momentum is 6 kgÅ"m/s due north. The rate of change of momentum is 3 (kgÅ"m/s)/s due north which is numerically equivalent to 3 newtons.

Conservation

In a closed system (one that does not exchange any matter with its surroundings and is not acted on by external forces) the total momentum is constant. This fact, known as the law of conservation of momentum, is implied by Newton's laws of motion. Suppose, for example, that two particles interact. Because of the third law, the forces between them are equal and opposite. If the particles are numbered 1 and 2, the second law states that F = dp/dt and F = dp/dt. Therefore,

$$\frac{dp_1}{dt} = -\frac{dp_2}{dt},$$

with the negative sign indicating that the forces oppose. Equivalently,

$$\frac{d}{dt}(p_1 + p_2) = 0.$$

If the velocities of the particles are u and u before the interaction, and afterwards they are v and v, then

$$m_1u_1 + m_2u_2 = m_1v_1 + m_2v_2.$$

This law holds no matter how complicated the force is between particles. Similarly, if there are several particles, the momentum exchanged between each pair of particles adds up to zero, so the total change in momentum is zero. This conservation law applies to all interactions, including collisions and separations caused by explosive forces. It can also be generalized to situations where Newton's laws do not hold, for example in the theory of relativity and in electrodynamics.

Dependence on reference frame

Momentum is a measurable quantity, and the measurement depends on the motion of the observer. For example: if an apple is sitting in a glass elevator that is descending, an outside observer, looking into the elevator, sees the apple moving, so, to that observer, the apple has a non-zero momentum.

To someone inside the elevator, the apple does not move, so, it has zero momentum. The two observers each have a frame of reference, in which, they observe motions, and, if the elevator is descending steadily, they will see behavior that is consistent with those same physical laws. Suppose a particle has position x in a stationary frame of reference. From the point of view of another frame of reference, moving at a uniform speed u, the position (represented by a primed coordinate) changes with time as

$$x' = x - ut.$$

This is called a Galilean transformation. If the particle is moving at speed dx/dt = v in the first frame of reference, in the second, it is moving at speed

$$v' = \frac{dx'}{dt} = v - u.$$

Since u does not change, the accelerations are the same:

$$a' = \frac{dv'}{dt} = a.$$

Thus, momentum is conserved in both reference frames. Moreover, as long as the force has the same form, in both frames, Newton's second law is unchanged. Forces such as Newtonian gravity, which depend only on the scalar distance between objects, satisfy this criterion. This independence of reference frame is called Newtonian relativity or Galilean invariance. A change of reference frame, can, often, simplify calculations of motion. For example, in a collision of two particles, a reference frame can be chosen, where, one particle begins at rest. Another, commonly used reference frame, is the center of mass frame – one that is moving with the center of mass. In this frame, the total momentum is zero.

Application to collisions

By itself, the law of conservation of momentum is not enough to determine the motion of particles after a collision. Another property of the motion, kinetic energy, must be known. This is not necessarily conserved. If it is conserved, the collision is called an elastic collision; if not, it is an inelastic collision.

Elastic collisions

Elastic collision of unequal masses

An elastic collision is one in which no kinetic energy is absorbed in the collision. Perfectly elastic "collisions" can occur when the objects do not touch each other, as for example in atomic or nuclear scattering where electric repulsion keeps them apart. A slingshot maneuver of a satellite around a planet can also be viewed as a perfectly elastic collision. A collision between two pool balls is a good example of an almost totally elastic collision, due to their high rigidity, but when bodies come in contact there is always some dissipation. A head-on elastic collision between two bodies can be represented by velocities in one dimension, along a line passing through the bodies. If the velocities are u and u before the collision and v and v after, the equations expressing conservation of momentum and kinetic energy are:

$$\begin{aligned} m_1u_1 + m_2u_2 &= m_1v_1 + m_2v_2 \\ \frac{1}{2}m_1u_1^2 + \frac{1}{2}m_2u_2^2 &= \frac{1}{2}m_1v_1^2 + \frac{1}{2}m_2v_2^2. \end{aligned}$$

A change of reference frame can simplify analysis of a collision. For example, suppose there are two bodies of equal mass m, one stationary and one approaching the other at a speed v (as in the figure). The center of mass is moving at speed v/2 and both bodies are moving towards it at speed v/2. Because of the symmetry, after the collision both must be moving away from the center of mass at the same speed. Adding the speed of the center of mass to both, we find that the body that was moving is now stopped and the other is moving away at speed v. The bodies have exchanged their velocities. Regardless of the velocities of the bodies, a switch to the center of mass frame leads us to the same conclusion. Therefore, the final velocities are given by

$$\begin{aligned} v_1 &= u_2 \\ v_2 &= u_1. \end{aligned}$$

In general, when the initial velocities are known, the final velocities are given by

$$v_1 = \left(\frac{m_1 - m_2}{m_1 + m_2}\right)u_1 + \left(\frac{2m_2}{m_1 + m_2}\right)u_2$$

$$v_2 = \left(\frac{m_2 - m_1}{m_1 + m_2}\right)u_2 + \left(\frac{2m_1}{m_1 + m_2}\right)u_1.$$

Inelastic collisions

In an inelastic collision, some of the kinetic energy of the colliding bodies is converted into other forms of energy (such as heat or sound). Examples include traffic collisions, in which the effect of loss of kinetic energy can be seen in the damage to the vehicles; electrons losing some of their energy to atoms (as in the Franck–Hertz experiment); and particle accelerators in which the kinetic energy is converted into mass in the form of new particles. In a perfectly inelastic collision (such as a bug hitting a windshield), both bodies have the same motion afterwards. If one body is motionless to begin with, the equation for conservation of momentum is

$$m_1u_1 = (m_1 + m_2)v,$$

So

$$v = \frac{m_1}{m_1 + m_2} u_1.$$

In a frame of reference moving at the speed v), the objects are brought to rest by the collision and 100% of the kinetic energy is converted to other forms of energy. One measure of the inelasticity of the collision is the coefficient of restitution C, defined as the ratio of relative velocity of separation to relative velocity of approach. In applying this measure to a ball bouncing from a solid surface, this can be easily measured using the following formula:

$$C_R = \sqrt{\frac{\text{bounce height}}{\text{drop height}}}.$$

The momentum and energy equations also apply to the motions of objects that begin together and then move apart. For example, an explosion is the result of a chain reaction that transforms potential energy stored in chemical, mechanical, or nuclear form into kinetic energy, acoustic energy, and electromagnetic radiation. Rockets also make use of conservation of momentum: propellant is thrust outward, gaining momentum, and an equal and opposite momentum is imparted to the rocket.

Multiple dimensions

Real motion has both direction and velocity and must be represented by a vector. In a coordinate system with x, y, z axes, velocity has components v in the x-direction, v in the y-direction, v in the z-direction. The vector is represented by a boldface symbol:

$$v = (c_x, v_y, v_z).$$

Similarly, the momentum is a vector quantity and is represented by a boldface symbol:

$$p = (p_x, p_y, p_z).$$

The equations in the previous sections, work in vector form if the scalars p and v are replaced by vectors p and v. Each vector equation represents three scalar equations. For example,

$$p_x = mv_x$$
$$p_y = mv_y$$
$$p_z = mv_z.$$

The kinetic energy equations are exceptions to the above replacement rule. The equations are still one-dimensional, but each scalar represents the magnitude of the vector, for example,

$$v^2 = v_x^2 + v_y^2 + v_z^2.$$

Each vector equation represents three scalar equations. Often coordinates can be chosen so that only two components are needed, as in the figure. Each component can be obtained separately and the results combined to produce a vector result. A simple construction involving the center of mass frame can be used to show that if a stationary elastic sphere is struck by a moving sphere, the two will head off at right angles after the collision (as in the figure).

Angular momentum

In physics, angular momentum (rarely, moment of momentum or rotational momentum) is the rotational equivalent of linear momentum. It is an important quantity in physics because it is a conserved quantity—the total angular momentum of a closed system remains constant.

In three dimensions, the angular momentum for a point particle is a pseudovector r × p, the cross product of the particle's position vector r (relative to some origin) and its momentum vector; the latter is p = mv in Newtonian mechanics. This definition can be applied to each point in continua like solids or fluids, or physical fields. Unlike momentum, angular momentum does depend on where the origin is chosen, since the particle's position is measured from it.

Just like for angular velocity, there are two special types of angular momentum: the spin angular momentum and the orbital angular momentum. The spin angular momentum of an object is defined as the angular momentum about its centre of mass coordinate. The orbital angular momentum of an object about a chosen origin is defined as the angular momentum of the centre of mass about the origin. The total angular momentum of an object is the sum of the spin and orbital angular momenta. The orbital angular momentum vector of a particle is always parallel and directly proportional to the orbital angular velocity vector ù of the particle, where the constant of proportionality depends on both the mass of the particle and its distance from origin. However, the spin angular momentum of the object is proportional but not always parallel to the spin angular velocity Ù, making the constant of proportionality a second-rank tensor rather than a scalar. Angular momentum is additive; the total angular momentum of any composite system is the (pseudo) vector sum of the angular momenta of its constituent parts. For a continuous rigid body, the total angular momentum is the volume integral of angular momentum density (i.e. angular momentum per unit volume in the limit as volume shrinks to zero) over the entire body.

Torque can be defined as the rate of change of angular momentum, analogous to force. The net external torque on any system is always equal to the total torque on the system; in other words, the sum of all internal torques of any system is always 0 (this is the rotational analogue of Newton's Third Law). Therefore, for a closed system (where there is no net external torque), the total torque on the system must be 0, which means that the total angular momentum of the system is constant. The conservation of angular momentum helps explain many observed phenomena, for example the increase in rotational speed of a spinning figure skater as the skater's arms are contracted, the high rotational rates of neutron stars, the Coriolis effect, and the precession of gyroscopes. In general, conservation does limit the possible motion of a system, but does not uniquely determine what the exact motion is.

In quantum mechanics, angular momentum (like other quantities) is expressed as an operator, and its one-dimensional projections have quantized eigenvalues. Angular momentum is subject to the Heisenberg uncertainty principle, implying that at any time, only one projection (also called "component") can be measured with definite precision; the other two then remain uncertain. Because of this, it turns out that the notion of a quantum particle literally "spinning" about an axis does not exist. Nevertheless, elementary particles still possess a spin angular momentum, but this angular momentum does not correspond to spinning motion in the ordinary sense.

In classical mechanics

Definition

Orbital angular momentum in two dimensions Angular momentum is a vector quantity (more precisely, a pseudovector) that represents the product of a body's rotational inertia and rotational velocity (in radians/sec) about a particular axis. However, if the particle's trajectory lies in a single plane, it is sufficient to discard the vector nature of angular momentum, and treat it as a scalar (more precisely, a pseudoscalar). Angular momentum can be considered a rotational analog of linear momentum. Thus, where linear momentum p is proportional to mass m and linear speed v

$$p = mv,$$

angular momentum L is proportional to moment of inertia I and angular speed ù measured in radians per second.

$$L = Iw.$$

Conservation of angular momentum

A rotational analog of Newton's third law of motion might be written, "In a closed system, no torque can be exerted on any matter without the exertion on some other matter of an equal and opposite torque." Hence, angular momentum can be exchanged between objects in a closed system, but total angular momentum before and after an exchange remains constant (is conserved). Seen another way, a rotational analogue of Newton's first law of motion might be written, "A rigid body continues in a state of uniform rotation unless acted by an external influence." Thus with no external influence to act upon it, the original angular momentum of the system remains constant.

The conservation of angular momentum is used in analyzing central force motion. If the net force on some body is directed always toward some point, the center, then there is no torque on the body with respect to the center, as all of the force is directed along the radius vector, and none is perpendicular to the radius. Mathematically, torque because in this case and are parallel vectors. Therefore, the angular momentum of the body about the center is constant. This is the case with gravitational attraction in the orbits of planets and satellites, where the gravitational force is always directed toward the primary body and orbiting bodies conserve angular momentum by exchanging distance and velocity as they move about the primary. Central force motion is also used in the analysis of the Bohr model of the atom.

For a planet, angular momentum is distributed between the spin of the planet and its revolution in its orbit, and these are often exchanged by various mechanisms. The conservation of angular momentum in the Earth-Moon system results in the transfer of angular momentum from Earth to Moon, due to tidal torque the Moon exerts on the Earth. This in turn results in the slowing down of the rotation rate of Earth, at about 65.7 nanoseconds per day, and in gradual increase of the radius of Moon's orbit, at about 3.82 centimeters per year. The conservation of angular momentum explains the angular acceleration of an ice skater as she brings her arms and legs close to the vertical axis of rotation. By bringing part of the mass of her body closer to the axis, she decreases her body's moment of inertia. Because angular momentum is the product of moment of inertia and angular velocity, if the angular momentum remains constant (is conserved), then the angular velocity (rotational speed) of the skater must increase.

The same phenomenon results in extremely fast spin of compact stars (like white dwarfs, neutron stars and black holes) when they are formed out of much larger and slower rotating stars. Decrease in the size of an object n times results in increase of its angular velocity by the factor of n. Conservation is not always a full explanation for the dynamics of a system but is a key constraint. For example, a spinning top is subject to gravitational torque making it lean over and change the angular momentum about the nutation axis, but neglecting friction at the point of spinning contact, it has a conserved angular momentum about its spinning axis, and another about its precession axis. Also, in any planetary system, the planets, star(s), comets, and asteroids can all move in numerous complicated ways, but only so that the angular momentum of the system is conserved.

Noether's theorem states that every conservation law is associated with a symmetry (invariant) of the underlying physics. The symmetry associated with conservation of angular momentum is rotational invariance. The fact that the physics of a system is unchanged if it is rotated by any angle about an axis implies that angular momentum is conserved.

In quantum mechanics

Angular momentum in quantum mechanics differs in many profound respects from angular momentum in classical mechanics. In relativistic quantum mechanics, it differs even more, in which the above relativistic definition becomes a tensorial operator.

Spin, orbital, and total angular momentum

- Right: extrinsic orbital angular momentum L about an axis.
- Top: the moment of inertia tensor I and angular velocity ? (L is not always parallel to ?).
- Bottom: momentum p and its radial position r from the axis. The total angular momentum (spin plus orbital) is J. For a quantum particle the interpretations are different; particle spin does not have the above interpretation.

The classical definition of angular momentum as can be carried over to quantum mechanics, by reinterpreting r as the quantum position operator and p as the quantum momentum operator. L is then an operator, specifically called the orbital angular momentum operator. The components of the angular momentum operator satisfy the commutation relations of the Lie algebra so(3). Indeed, these operators are precisely the infinitesimal action of the rotation group on the quantum Hilbert space. (the discussion below of the angular momentum operators as the generators of rotations.)

However, in quantum physics, there is another type of angular momentum, called spin angular momentum, represented by the spin operator S. Almost all elementary particles have spin. Spin is often depicted as a particle literally spinning around an axis, but this is a misleading and inaccurate picture: spin is an intrinsic property of a particle, unrelated to any sort of motion in space and fundamentally different from orbital angular momentum. All elementary particles have a characteristic spin, for example electrons have "spin (this actually means "spin while photons have "spin 1".

Finally, there is total angular momentum J, which combines both the spin and orbital angular momentum of all particles and fields. (For one particle, J = L + S.) Conservation of angular momentum applies to J, but not to L or S; for example, the spin-orbit interaction allows angular momentum to transfer back and forth between L and S, with the total remaining constant. Electrons and photons need not have integer-based values for total angular momentum, but can also have fractional values.

History

Newton, in the Principia, hinted at angular momentum in his examples of the First Law of Motion,

A top, whose parts by their cohesion are perpetually drawn aside from rectilinear motions, does not cease its rotation, otherwise than as it is retarded by the air. The greater bodies of the planets and comets, meeting with less resistance in more free spaces, preserve their motions both progressive and circular for a much longer time. He did not further investigate angular momentum directly in the Principia, From such kind of reflexions also sometimes arise the circular motions of bodies about their own centres. But these are cases which I do not consider in what follows; and it would be too tedious to demonstrate every particular that relates to this subject. However, his geometric proof of the law of areas is an outstanding example of Newton's genius, and indirectly proves angular momentum conservation in the case of a central force.

The Law of Areas

As a planet orbits the Sun, the line between the Sun and the planet sweeps out equal areas in equal intervals of time. This had been known since Kepler expounded his second law of planetary motion. Newton derived a unique geometric proof, and went on to show that the attractive force of the Sun's gravity was the cause of all of Kepler's laws.

During the first interval of time, an object is in motion from point A to point B. Undisturbed, it would continue to point c during the second interval. When the object arrives at B, it receives an impulse directed toward point S. The impulse gives it a small added velocity toward S, such that if this were its only velocity, it would move from B to V during the second interval. By the rules of velocity composition, these two velocities add, and point C is found by construction of parallelogram BcCV. Thus the object's path is deflected by the impulse so that it arrives at point C at the end of the second interval. Because the triangles SBc and SBC have the same base SB and the same height Bc or VC, they have the same area. By symmetry, triangle SBc also has the same area as triangle SAB, therefore the object has swept out equal areas SAB and SBC in equal times.

At point C, the object receives another impulse toward S, again deflecting its path during the third interval from d to D. Thus it continues to E and beyond, the triangles SAB, SBc, SBC, SCd, SCD, SDe, SDE all having the same area. Allowing the time intervals to become ever smaller, the path ABCDE approaches indefinitely close to a continuous curve.

Note that because this derivation is geometric, and no specific force is applied, it proves a more general law than Kepler's second law of planetary motion. It shows that the Law of Areas applies to any central force, attractive or repulsive, continuous or non-continuous, or zero.

Conservation of angular momentum in the Law of Areas

The proportionality of angular momentum to the area swept out by a moving object can be understood by realizing that the bases of the triangles, that is, the lines from S to the object, are equivalent to the radius r, and that the heights of the triangles are proportional to the perpendicular component of velocity v. Hence, if the area swept per unit time is constant, then by the triangular area formula 1/2(base)(height), the product (base)(height) and therefore the product rv are constant: if r and the base length are decreased, v and height must increase proportionally. Mass is constant, therefore angular momentum rmv is conserved by this exchange of distance and velocity. In the case of triangle SBC, area is equal to 1/2(SB)(VC). Wherever C is eventually located due to the impulse applied at B, the product (SB)(VC), and therefore rmv remain constant. Similarly so for each of the triangles.

After Newton

Leonhard Euler, Daniel Bernoulli, and Patrick d'Arcy all understood angular momentum in terms of conservation of areal velocity, a result of their analysis of Kepler's second law of planetary motion. It is unlikely that they realized the implications for ordinary rotating matter. In 1736 Euler, like Newton, touched on some of the equations of angular momentum in his Mechanica without further developing them. Bernoulli wrote in a 1744 letter of a "moment of rotational motion", possibly the first conception of angular momentum as we now understand it. In 1799, Pierre-Simon Laplace first realized that a fixed plane was associated with rotation - his invariable plane. Louis Poinsot in 1803 began representing rotations as a line segment perpendicular to the rotation, and elaborated on the "conservation of moments". In 1852 Léon Foucault used a gyroscope in an experiment to display the Earth's rotation. William J. M. Rankine's 1858 Manual of Applied Mechanics defined angular momentum in the modern sense for the first time:...a line whose length is proportional to the magnitude of the angular momentum, and whose direction is perpendicular to the plane of motion of the body and of the fixed point, and such, that when the motion of the body is viewed from the extremity of the line, the radius-vector of the body seems to have right-handed rotation.

In an 1872 edition of the same book, Rankine stated that "The term angular momentum was introduced by Mr. Hayward," probably referring to R.B. Hayward's article On a Direct Method of estimating Velocities, Accelerations, and all similar Quantities with respect to Axes moveable in any manner in Space with Applications, which was introduced in 1856, and published in 1864. Rankine was mistaken, as numerous publications feature the term starting in the late 18th to early 19th centuries. However, Hayward's article apparently was the first use of the term and the concept seen by much of the English-speaking world. Before this, angular momentum was typically referred to as "momentum of rotation" in English.

Collision

A collision is the event in which two or more bodies exert forces on each other in about a relatively short time. Although the most common use of the word collision refers to incidents in which two or more objects collide with great force, the scientific use of the term implies nothing about the magnitude of the force. Some examples of physical interactions that scientists would consider collisions are the following:

- When an insect lands on a plant's leaf, its legs are said to collide with the leaf.
- When a cat strides across a lawn, each contact that its paws make with the ground is considered a collision, as well as each brush of its fur against a blade of grass.
- When a boxer throws a punch, his fist is said to collide with the opponent's body.

- When an astronomical object merges with a black hole, they are considered to collide.

Some colloquial uses of the word collision are the following:

- A traffic collision involves at least one automobile.
- A mid-air collision occurs between airplanes.
- A ship collision accurately involves at least two moving maritime vessels hitting each other, or the related term, allision, when a moving ship strikes a stationary ship. Overview Collision is short-duration interaction between two bodies or more than two bodies simultaneously causing change in motion of bodies involved due to internal forces acted between them during this. Collisions involve forces (there is a change in velocity). The magnitude of the velocity difference just before impact is called the closing speed. All collisions conserve momentum. What distinguishes different types of collisions is whether they also conserve kinetic energy. The Line of impact is the line that is colinear to the common normal of the surfaces that are closest or in contact during impact. This is the line along which internal force of collision acts during impact, and Newton's coefficient of restitution is defined only along this line. Collisions are of three types:

1. Perfectly elastic collision
2. Inelastic collision
3. Perfectly inelastic collision.

Specifically, collisions can either be elastic, meaning they conserve both momentum and kinetic energy, or inelastic, meaning they conserve momentum but not kinetic energy. An inelastic collision is sometimes also called a plastic collision. A "perfectly inelastic" collision (also called a "perfectly plastic" collision) is a limiting case of inelastic collision in which the two bodies coalesce after impact. The degree to which a collision is elastic or inelastic is quantified by the coefficient of restitution, a value that generally ranges between zero and one. A perfectly elastic collision has a coefficient of restitution of one; a perfectly inelastic collision has a coefficient of restitution of zero.

TYPES OF COLLISIONS

There are two types of collisions between two bodies - 1) Head-on collisions or one-dimensional collisions - where the velocity of each body just before impact is along the line of impact, and 2) Non-head-on collisions, oblique collisions or two-dimensional collisions - where the velocity of each body just before impact is not along the line of impact. According to the coefficient of restitution, there are two special cases of any collision as written below:

1. A perfectly elastic collision is defined as one in which there is no loss of kinetic energy in the collision. In reality, any macroscopic collision between objects will convert some kinetic energy to internal energy and other forms of energy, so no large-scale impacts are perfectly elastic. However, some problems are sufficiently close to perfectly elastic that they can be approximated as such. In this case, the coefficient of restitution equals one.
2. An inelastic collision is one in which part of the kinetic energy is changed to some other form of energy in the collision. Momentum is conserved in inelastic collisions (as it is for elastic collisions), but one cannot track the kinetic energy through the collision since some of it is converted to other forms of energy. In this case, coefficient of restitution does not equal one.

In any type of collision there is a phase when for a moment colliding bodies have the same velocity along the line of impact.Then the kinetic energy of bodies reduces to its minimum during this phase and may be called a maximum deformation phase for which momentarily the coefficient of restitution becomes one. Collisions in ideal gases approach perfectly elastic collisions, as do scattering interactions of sub-atomic particles which are deflected by the electromagnetic force. Some large-scale interactions like the slingshot type gravitational interactions between satellites and planets are perfectly elastic. Collisions between hard spheres may be nearly elastic, so it is useful to calculate the limiting case of an elastic collision. The assumption of conservation of momentum as well as the conservation of kinetic energy makes possible the calculation of the final velocities in two-body collisions.

Allision

In maritime law, it is occasionally desirable to distinguish between the situation of a vessel striking a moving object, and that of it striking a stationary object. The word "allision" is then used to mean the striking of a stationary object, while "collision" is used to mean the striking of a moving object. So when two vessels run against each other, it is called collision whereas when one vessel ran against another, it is considered allision. The fixed object could also include a bridge or dock. While there is no huge difference between the two terminologies and often they are even used interchangeably, it is important to determine the difference because it helps clarify the circumstances of emergencies and adapt accordingly. In the case of Vane Line Bunkering, Inc. v. Natalie D M/V, it was established that there was the presumption that the moving vessel is at fault, stating that "presumption derives from the common-sense observation that moving vessels do not usually collide with stationary objects unless the [moving] vessel is mishandled in some way." This is also referred to as The Oregon Rule.

Analytical vs. numerical approaches towards resolving collisions Relatively few problems involving collisions can be solved analytically; the remainder require numerical methods. An important problem in simulating collisions is determining whether two objects have in fact collided. This problem is called collision detection Examples of collisions that can be solved analytically

Billiards

Collisions play an important role in cue sports. Because the collisions between billiard balls are nearly elastic, and the balls roll on a surface that produces low rolling friction, their behavior is often used to illustrate Newton's laws of motion. After a zero-friction collision of a moving ball with a stationary one of equal mass, the angle between the directions of the two balls is 90 degrees. This is an important fact that professional billiards players take into account, although it assumes the ball is moving frictionlessly across the table rather than rolling with friction. Consider an elastic collision in 2 dimensions of any 2 masses m and m, with respective initial velocities u and u where u = 0, and final velocities V and V. Conservation of momentum gives mu = mV+ mV. Conservation of energy for an elastic collision gives (1/2)m|u| = (1/2)m|V| + (1/2)m|V|. Now consider the case m = m: we obtain u=V+V and |u| = |V|+|V|. Taking the dot product of each side of the former equation with itself, |u| = uou = |V|+|V|+2VoV. Comparing this with the latter equation gives VoV = 0, so they are perpendicular unless V is the zero vector (which occurs if and only if the collision is head-on).

Elastic collision

An elastic collision is an encounter between two bodies in which the total kinetic energy of the two bodies remains the same. In an ideal, perfectly elastic collision, there is no net conversion of kinetic energy into other forms such as heat, noise, or potential energy. During the collision of small objects, kinetic energy is first converted to potential energy associated with a repulsive force between the particles (when the particles move against this force, i.e. the angle between the force and the relative velocity is obtuse), then this potential energy is converted back to kinetic energy (when the particles move with this force, i.e. the angle between the force and the relative velocity is acute).

Collisions of atoms are elastic, for example Rutherford backscattering. A useful special case of elastic collision is when the two bodies have equal mass, in which case they will simply exchange their momenta. The molecules-as distinct from atoms-of a gas or liquid rarely experience perfectly elastic collisions because kinetic energy is exchanged between the molecules' translational motion and their internal degrees of freedom with each collision. At any instant, half the collisions are, to a varying extent, inelastic collisions (the pair possesses less kinetic energy in their translational motions after the collision than before), and half could be described as "super-elastic" (possessing more kinetic energy after the collision than before). Averaged across the entire sample, molecular collisions can be regarded as essentially elastic as long as Planck's law forbids black-body photons to carry away energy from the system. In the case of macroscopic bodies, perfectly elastic collisions are an ideal never fully realized, but approximated by the interactions of objects such as billiard balls. When considering energies, possible rotational energy before and/or after a collision may also play a role.

Two-dimensional

For the case of two colliding bodies in two dimensions, the overall velocity of each body must be split into two perpendicular velocities: one tangent to the common normal surfaces of the colliding bodies at the point of contact, the other along the line of collision. Since the collision only imparts force along the line of collision, the velocities that are tangent to the point of collision do not change. The velocities along the line of collision can then be used in the same equations as a one-dimensional collision. The final velocities can then be calculated from the two new component velocities and will depend on the point of collision. Studies of two-dimensional collisions are conducted for many bodies in the framework of a two-dimensional gas.

In a center of momentum frame at any time the velocities of the two bodies are in opposite directions, with magnitudes inversely proportional to the masses. In an elastic collision these magnitudes do not change. The directions may change depending on the shapes of the bodies and the point of impact. For example, in the case of spheres the angle depends on the distance between the (parallel) paths of the centers of the two bodies. Any non-zero change of direction is possible: if this distance is zero the velocities are reversed in the collision; if it is close to the sum of the radii of the spheres the two bodies are only slightly deflected.

Inelastic collision

An inelastic collision, in contrast to an elastic collision, is a collision in which kinetic energy is not conserved due to the action of internal friction. In collisions of macroscopic bodies, some kinetic energy is turned into vibrational energy of the atoms, causing a heating effect, and the bodies are deformed. The molecules of a gas or liquid rarely experience perfectly elastic collisions because kinetic energy is exchanged between the molecules' translational motion and their internal degrees of freedom with each collision. At any one instant, half the collisions are - to a varying extent - inelastic (the pair possesses less kinetic energy after the collision than before), and half could be described as "super-elastic" (possessing more kinetic energy after the collision than before). Averaged across an entire sample, molecular collisions are elastic. Although inelastic collisions do not conserve kinetic energy, they do obey conservation of momentum. Simple ballistic pendulum problems obey the conservation of kinetic energy only when the block swings to its largest angle.

In nuclear physics, an inelastic collision is one in which the incoming particle causes the nucleus it strikes to become excited or to break up. Deep inelastic scattering is a method of probing the structure of subatomic particles in much the same way as Rutherford probed the inside of the atom (see Rutherford scattering). Such experiments were performed on protons in the late 1960s using high-energy electrons at the Stanford Linear Accelerator (SLAC). As in Rutherford scattering, deep inelastic scattering of electrons by proton targets revealed that most of the incident electrons interact very little and pass straight through, with only a small number bouncing back. This indicates that the charge in the proton is concentrated in small lumps, reminiscent of Rutherford's discovery that the positive charge in an atom is concentrated at the nucleus. However, in the case of the proton, the evidence suggested three distinct concentrations of charge (quarks) and not one.

Center of mass

In physics, the center of mass of a distribution of mass in space is the unique point where the weighted relative position of the distributed mass sums to zero. This is the point to which a force may be applied to cause a linear acceleration without an angular acceleration. Calculations in mechanics are often simplified when formulated with respect to the center of mass. It is a hypothetical point where entire mass of an object may be assumed to be concentrated to visualise its motion. In other words, the center of mass is the particle equivalent of a given object for application of Newton's laws of motion.

In the case of a single rigid body, the center of mass is fixed in relation to the body, and if the body has uniform density, it will be located at the centroid. The center of mass may be located outside the physical body, as is sometimes the case for hollow or open-shaped objects, such as a horseshoe. In the case of a distribution of separate bodies, such as the planets of the Solar System, the center of mass may not correspond to the position of any individual member of the system.

The center of mass is a useful reference point for calculations in mechanics that involve masses distributed in space, such as the linear and angular momentum of planetary bodies and rigid body dynamics. In orbital mechanics, the equations of motion of planets are formulated as point masses located at the centers of mass. The center of mass frame is an inertial frame in which the center of mass of a system is at rest with respect to the origin of the coordinate system.

History

The concept of "center of mass" in the form of the center of gravity was first introduced by the great ancient Greek physicist, mathematician, and engineer Archimedes of Syracuse. He worked with simplified assumptions about gravity that amount to a uniform field, thus arriving at the mathematical properties of what we now call the center of mass. Archimedes showed that the torque exerted on a lever by weights resting at various points along the lever is the same as what it would be if all of the weights were moved to a single point-their center of mass. In work on floating bodies he demonstrated that the orientation of a floating object is the one that makes its center of mass as low as possible. He developed mathematical techniques for finding the centers of mass of objects of uniform density of various well-defined shapes.

Later mathematicians who developed the theory of the center of mass include Pappus of Alexandria, Guido Ubaldi, Francesco Maurolico, Federico Commandino, Simon Stevin, Luca Valerio, Jean-Charles de la Faille, Paul Guldin, John Wallis, Louis Carré, Pierre Varignon, and Alexis Clairaut. Newton's second law is reformulated with respect to the center of mass in Euler's first law.

Definition

The center of mass is the unique point at the center of a distribution of mass in space that has the property that the weighted position vectors relative to this point sum to zero. In analogy to statistics, the center of mass is the mean location of a distribution of mass in space.

A system of particles

In the case of a system of particles P, i = 1,n…,n, each with mass m that are located in space with coordinates r, i = 1,n…,n, the coordinates R of the center of mass satisfy the condition

$$\sum_{i=1}^{n} m_i(\mathbf{r}_i - \mathbf{R}) = 0.$$

Solving this equation for R yields the formula

$$\mathbf{R} = \frac{1}{M}\sum_{i=1}^{n} m_i \mathbf{r}_i,$$

where M is the sum of the masses of all of the particles.

A continuous volume

If the mass distribution is continuous with the density ñ(r) within a solid Q, then the integral of the weighted position coordinates of the points in this volume relative to the center of mass R over the volume V is zero, that is

$$\iiint_Q \rho(\mathbf{r})(\mathbf{r} - \mathbf{R})dV = 0.$$

Solve this equation for the coordinates R to obtain

$$\mathbf{R} = \frac{1}{M}\iiint_Q \rho(\mathbf{r})\mathbf{r}dV,$$

where M is the total mass in the volume.

If a continuous mass distribution has uniform density, which means ñ is constant, then the center of mass is the same as the centroid of the volume.

6

HEAT

In physics, heat is energy in transfer other than as work or by transfer of matter. When there is a suitable physical pathway, heat flows from a hotter body to a colder one. The transfer results in a net increase in entropy. The pathway can be direct, as in conduction and radiation, or indirect, as in convective circulation.

Heat refers to a process of transfer between two systems, the system of interest, and its surroundings considered as a system, not to a state or property of a single system. If heat transfer is slow and continuous, so that the temperature of the system of interest remains well defined, it can be described by a process function.

Kinetic theory explains heat as a macroscopic manifestation of the motions and interactions of microscopic constituents such as molecules and photons.

In calorimetry, sensible heat is defined with respect to a particular state variable of the system; it causes change of temperature, leaving that particular state variable unchanged. Heat transfer that occurs with the system at constant temperature and that changes that particular state variable is called latent heat with respect to that variable. For infinitesimal changes, the total incremental heat transfer is then the sum of the latent and sensible heat increments. This is a basic paradigm for thermodynamics, and was important in the historical development of the subject.

The quantity of energy transferred as heat is a scalar expressed in an energy unit such as the joule (J) (SI), with a sign that is customarily positive when a transfer adds to the energy of a system. It can be measured by calorimetry, or determined by calculations based on other quantities, relying on the first law of thermodynamics.

HISTORY

Physicist James Clerk Maxwell, in his 1871 classic Theory of Heat, was one of many who began to build on the already established idea that heat has something to do with matter in motion. This was the same idea put forth by Benjamin Thompson in 1798, who said he was only following up on the work of many others. One of Maxwell's recommended books was Heat as a Mode of Motion, by John Tyndall. Maxwell outlined four stipulations for the definition of heat:

- It is something which may be transferred from one body to another, according to the second law of thermodynamics.
- It is a measurable quantity, and so can be treated mathematically.
- It cannot be treated as a material substance, because it may be transformed into something that is not a material substance, e. g., mechanical work.
- Heat is one of the forms of energy.

From empirically based ideas of heat, and from other empirical observations, the notions of internal energy and of entropy can be derived, so as to lead to the recognition of the first and second laws of thermodynamics. This was the way of the historical pioneers of thermodynamics.

TRANSFERS OF ENERGY AS HEAT BETWEEN TWO BODIES

Referring to conduction, Partington writes: "If a hot body is brought in conducting contact with a cold body, the temperature of the hot body falls and that of the cold body rises, and it is said that a quantity of heat has passed from the hot body to the cold body. "

Referring to radiation, Maxwell writes: "In Radiation, the hotter body loses heat, and the colder body receives heat by means of a process occurring in some intervening medium which does not itself thereby become hot. "

Maxwell writes that convection as such "is not a purely thermal phenomenon". In thermodynamics, convection in general is regarded as transport of internal energy. If, however, the convection is enclosed and circulatory, then it may be regarded as an intermediary that transfers energy as heat between source and destination bodies, because it transfers only energy and not matter from the source to the destination body.

Practical operating devices that harness transfers of energy as heat In accordance with the first law for closed systems, energy transferred solely as heat enters one body and leaves another, changing the internal energies of each. Transfer, between bodies, of energy as work is a complementary way of changing internal energies. Though it is not logically rigorous from the viewpoint of strict physical concepts, a common form of words that expresses this is to say that heat and work are interconvertible.

Heat engine

In classical thermodynamics, a commonly considered model is the heat engine. It consists of four bodies: the working body, the hot reservoir, the cold reservoir, and the work reservoir. A cyclic process leaves the working body in an unchanged state, and is envisaged as being repeated indefinitely often. Work transfers between the working body and the work reservoir are envisaged as reversible, and thus only one work reservoir is needed. But two thermal reservoirs are needed, because transfer of energy as heat is irreversible. A single cycle sees energy taken by the working body from the hot reservoir and sent to the two other reservoirs, the work reservoir and the cold reservoir. The hot reservoir always and only supplies energy and the cold reservoir always and only receives energy. The second law of thermodynamics requires that no cycle can occur in which no energy is received by the cold reservoir. Heat engines achieve higher efficiency when the difference between initial and final temperature is greater.

Heat pump

Another commonly considered model is the heat pump or refrigerator. Again there are four bodies: the working body, the hot reservoir, the cold reservoir, and the work reservoir. A single cycle starts with the working body colder than the cold reservoir, and then energy is taken in as heat by the working body from the cold reservoir. Then the work reservoir does work on the working body, adding more to its internal energy, making it hotter than the hot reservoir. The hot working body passes heat to the hot reservoir, but still remains hotter than the cold reservoir. Then, by allowing it to expand without doing work on another body and without passing heat to another body, the working body is made colder than the cold reservoir. It can now accept heat transfer from the cold reservoir to start another cycle. The device has transported energy from a colder to a hotter reservoir, but this is not regarded as being by an inanimate agency. This is because work is supplied from the work reservoir, not just by a simple thermodynamic process, but by a sequence ofthermodynamic operations, which may be regarded as directed by an animate agency. Accordingly, the cycle is still in accord with the second law of thermodynamics. The efficiency of a heat pump is best when the temperature difference between the hot and cold reservoirs is least.

Macroscopic view of quantity of energy transferred as heat According to Planck, there are three main conceptual approaches to heat. One is the microscopic or kinetic theory approach. Also there are two macroscopic approaches. One is the approach through the law of conservation of energy taken as prior to thermodynamics, with a mechanical analysis of processes, for example in the work of Helmholtz. This mechanical view is taken as currently customary in this article. The other macroscopic approach is the thermodynamic one, which admits heat as a primitive concept, which contributes, by scientific induction to knowledge of the law of conservation of energy.

Bailyn also distinguishes the two macroscopic approaches as the mechanical and the thermodynamic. The thermodynamic view was taken by the founders of thermodynamics in the nineteenth century. It regards quantity of energy transferred as heat as a primitive concept coherent with a primitive concept of temperature, measured primarily by calorimetry. A calorimeter is a body in the surroundings of the system, with its own temperature and internal energy; when it is connected to the system by a path for heat transfer, changes in it measure heat transfer. The mechanical view was pioneered by Helmholtz and developed and used in the twentieth century, largely through the influence of Max Born. It regards quantity of heat transferred as heat as a derived concept, defined for closed systems as quantity of heat transferred by mechanisms other than work transfer, the latter being regarded as primitive for thermodynamics, defined by macroscopic mechanics. According to Born, the transfer of internal energy between open systems that accompanies transfer of matter "cannot be reduced to mechanics". It follows that there is no well-founded definition of quantities of energy transferred as heat or as work associated with transfer of matter.

Nevertheless, for the thermodynamical description of non-equilibrium processes, it is desired to consider the effect of a temperature gradient established by the surroundings across the system of interest when there is no physical barrier or wall between system and surroundings, that is to say, when they are open with respect to one another. The impossibility of a mechanical definition in terms of work for this circumstance does not alter the physical fact that a temperature gradient causes a diffusive flux of internal energy, a process that, in the thermodynamic view, might be proposed as a candidate concept for transfer of energy as heat.

In this circumstance, it may be expected that there may also be active other drivers of diffusive flux of internal energy, such as gradient of chemical potential which drives transfer of matter, and gradient of electric potential which drives electric current and iontophoresis; such effects usually interact with diffusive flux of internal energy driven by temperature gradient, and such interactions are known as cross-effects.

If cross-effects that result in diffusive transfer of internal energy were also labeled as heat transfers, they would sometimes violate the rule that pure heat transfer occurs only down a temperature gradient, never up one. They would also contradict the principle that all heat transfer is of one and the same kind, a principle founded on the idea of heat conduction between closed systems. One might to try to think narrowly of heat flux driven purely by temperature gradient as a conceptual component of diffusive internal energy flux, in the thermodynamic view, the concept resting specifically on careful calculations based on detailed knowledge of the processes and being indirectly assessed. In these circumstances, if perchance it happens that no transfer of matter is actualized, and there are no cross-effects, then the thermodynamic concept and the mechanical concept coincide, as if one were dealing with closed systems. But when there is transfer of matter, the exact laws by which temperature gradient drives diffusive flux of internal energy, rather than being exactly knowable, mostly need to be assumed, and in many cases are practically unverifiable. Consequently, when there is transfer of matter, the calculation of the pure 'heat flux' component of the diffusive flux of internal energy rests on practically unverifiable assumptions. This is a reason to think of heat as a specialized concept that relates primarily and precisely to closed systems, and applicable only in a very restricted way to open systems.

In many writings in this context, the term "heat flux" is used when what is meant is therefore more accurately called diffusive flux of internal energy; such usage of the term "heat flux" is a residue of older and now obsolete language usage that allowed that a body may have a "heat content".

Microscopic view of heat

In the kinetic theory, heat is explained in terms of the microscopic motions and interactions of constituent particles, such as electrons, atoms, and molecules. Heat transfer arises from temperature gradients or differences, through the diffuse exchange of microscopic kinetic and potential particle energy, by particle collisions and other interactions. An early and vague expression of this was made by Francis Bacon. Precise and detailed versions of it were developed in the nineteenth century.

In statistical mechanics, for a closed system (no transfer of matter), heat is the energy transfer associated with a disordered, microscopic action on the system, associated with jumps in occupation numbers of the energy levels of the system, without change in the values of the energy levels themselves. It is possible for macroscopic thermodynamic work to alter the occupation numbers without change in the values of the system energy levels themselves, but what distinguishes transfer as heat is that the transfer is entirely due to disordered, microscopic action, including radiative transfer. A mathematical definition can be formulated for small increments of quasi-static adiabatic work in terms of the statistical distribution of an ensemble of microstates.

Notation and units

As a form of energy heat has the unit joule (J) in the International System of Units (SI). However, in many applied fields in engineering the British thermal unit (BTU) and the calorieare often used. The standard unit for the rate of heat transferred is the watt (W), defined as joules per second.

The total amount of energy transferred as heat is conventionally written as Q (from Quantity) for algebraic purposes. Heat released by a system into its surroundings is by convention a negative quantity ($Q < 0$); when a system absorbs heat from its surroundings, it is positive ($Q > 0$). Heat transfer rate, or heat flow per unit time, is denoted by. This should not be confused with a time derivative of a function of state (which can also be written with the dot notation) since heat is not a function of state. Heat flux is defined as rate of heat transfer per unit cross-sectional area, resulting in the unit watts per square metre.

ESTIMATION OF QUANTITY OF HEAT

Quantity of heat transferred can measured by calorimetry, or determined through calculations based on other quantities.

Calorimetry is the empirical basis of the idea of quantity of heat transferred in a process. The transferred heat is measured by changes in a body of known properties, for example, temperature rise, change in volume or length, or phase change, such as melting of ice.

A calculation of quantity of heat transferred can rely on a hypothetical quantity of energy transferred as adiabatic work and on the first law of thermodynamics. Such calculation is the primary approach of many theoretical studies of quantity of heat transferred.

Internal energy and enthalpy

For a closed system (a system from which no matter can enter or exit), one version of the first law of thermodynamics states that the change in internal energy ÄU of the system is equal to the amount of heat Q supplied to the system minus the amount of work W done by system on its surroundings. The foregoing sign convention for work is used in the present article, but an alternate sign convention, followed by IUPAC, for work, is to consider the work performed on the system by its surroundings as positive. This is the convention adopted by many modern textbooks of physical chemistry, such as those by Peter Atkins and Ira Levine, but many textbooks on physics define work as work done by the system.

This formula can be re-written so as to express a definition of quantity of energy transferred as heat, based purely on the concept of adiabatic work, if it is supposed that ÄU is defined and measured solely by processes of adiabatic work:

The work done by the system includes boundary work (when the system increases its volume against an external force, such as that exerted by a piston) and other work (e. g. shaft work performed by a compressor fan), which is called isochoric work:

In this Section we will neglect the "other-" or isochoric work contribution.

The internal energy, U, is a state function. In cyclical processes, such as the operation of a heat engine, state functions of the working substance return to their initial values upon completion of a cycle.

The differential, or infinitesimal increment, for the internal energy in an infinitesimal process is an exact differential dU. The symbol for exact differentials is the lowercase letter d.

In contrast, neither of the infinitesimal increments äQ nor äW in an infinitesimal process represents the state of the system. Thus, infinitesimal increments of heat and work are inexact differentials. The lowercase Greek letter delta, ä, is the symbol for inexact differentials. The integral of any inexact differential over the time it takes for a system to leave and return to the same thermodynamic state does not necessarily equal zero.

As recounted below, in the section headed Entropy, the second law of thermodynamics observes that if heat is supplied to a system in which no irreversible processes take place and which has a well-defined temperature T, the increment of heat äQ and the temperature T form the exact differential

$$\mathrm{d}S = \frac{\delta Q}{T},$$

and that S, the entropy of the working body, is a function of state. Likewise, with a well-defined pressure, P, behind the moving boundary, the work differential, äW, and the pressure, P, combine to form the exact differential

$$dV = \frac{\delta W}{P},$$

with V the volume of the system, which is a state variable. In general, for homogeneous systems,dU = TdS - PdV.

Associated with this differential equation is that the internal energy may be considered to be a function U (S, V) of its natural variables S and V. The internal energy representation of the fundamental thermodynamic relation is written U = U(S,V)

If V is constant TdS = dU (V constant)

and if P is constant TdS = dH (P constant)

with H the enthalpy defined byH = U + PV

ENTROPY

In 1856, German physicist Rudolf Clausius, referring to closed systems, in which transfer of matter does not occur, defined the second fundamental theorem (the second law of thermodynamics) in the mechanical theory of heat (thermodynamics): "if two transformations which, without necessitating any other permanent change, can mutually replace one another, be called equivalent, then the generations of the quantity of heat Q from work at the temperature T, has the equivalence-value:"In 1865, he came to define the entropy symbolized by S, such that, due to the supply of the amount of heat Q at temperature T the entropy of the system is increased by

$$\Delta S = \frac{Q}{T} \qquad (1)$$

In a transfer of energy as heat without work being done, there are changes of entropy in both the surroundings which lose heat and the system which gains it. The increase, ÄS, of entropy in the system may be considered to consist of two parts, an increment, ÄS2 that matches, or 'compensates', the change, "ÄS2, of entropy in the surroundings, and a further increment, ÄS2 2 that may be considered to be 'generated' or 'produced' in the system, and is said therefore to be 'uncompensated'. Thus

It is then said that an amount of entropy ÄS2 has been transferred from the surroundings to the system. Because entropy is not a conserved quantity, this is an exception to the general way of speaking, in which an amount transferred is of a conserved quantity.

The second law of thermodynamics observes that in a natural transfer of energy as heat, in which the temperature of the system is different from that of the surroundings, it is always so that $\Delta S_{overall} > 0$

For purposes of mathematical analysis of transfers, one thinks of fictive processes that are called 'reversible', with the temperature T of the system being hardly less than that of the surroundings, and the transfer taking place at an imperceptibly slow speed.

Following the definition above in formula (1), for such a fictive 'reversible' process, a quantity of transferred heat äQ (an inexact differential) is analyzed as a quantity T dS, with dS(an exact differential):TdS - δQ

This equality is only valid for a fictive transfer in which there is no production of entropy, that is to say, in which there is no uncompensated entropy.

which is the second law of thermodynamics for closed systems.

In non-equilibrium thermodynamics that approximates by assuming the hypothesis of local thermodynamic equilibrium, there is a special notation for this. The transfer of energy as heat is assumed to take place across an infinitesimal temperature difference, so that the system element and its surroundings have near enough the same temperature T.

Latent and sensible heat

In an 1847 lecture entitled On Matter, Living Force, and Heat, James Prescott Joule characterized the terms latent heat and sensible heat as components of heat each affecting distinct physical phenomena, namely the potential and kinetic energy of particles, respectively. He described latent energy as the energy possessed via a distancing of particles where attraction was over a greater distance, i. e. a form of potential energy, and the sensible heat as an energy involving the motion of particles or what was known as a living force. At the time of Joule kinetic energy either held 'invisibly' internally or held 'visibly' externally was known as a living force.

Latent heat is the heat released or absorbed by a chemical substance or a thermodynamic system during a change of state that occurs without a change in temperature. Such a process may be a phase transition, such as the melting of ice or the boiling of water. The term was introduced around 1750 byJoseph Black as derived from the Latin latere (to lie hidden), characterizing its effect as not being directly measurable with a thermometer.

Sensible heat, in contrast to latent heat, is the heat transferred to a thermodynamic system that has as its sole effect a change of temperature.

Both latent heat and sensible heat transfers increase the internal energy of the system to which they are transferred.

Consequences of Black's distinction between sensible and latent heat are examined in the Wikipedia article on calorimetry.

Specific heat

Specific heat, also called specific heat capacity, is defined as the amount of energy that has to be transferred to or from one unit of mass (kilogram) or amount of substance (mole) to change the system temperature by one degree. Specific heat is a physical property, which means that it depends on the substance under consideration and its state as specified by its properties.

The specific heats of monatomic gases (e. g., helium) are nearly constant with temperature. Diatomic gases such as hydrogen display some temperature dependence, and triatomic gases (e. g., carbon dioxide) still more.

Relation between heat, hotness, and temperature

According to Baierlein, a system's hotness is its tendency to transfer energy as heat. All physical systems are capable of heating or cooling others. This does not require that they have thermodynamic temperatures. With reference to hotness, the comparative terms hotter and colder are defined by the rule that heat flows from the hotter body to the colder.

If a physical system is inhomogeneous or very rapidly or irregularly changing, for example by turbulence, it may be impossible to characterize it by a temperature, but still there can be transfer of energy as heat between it and another system. If a system has a physical state that is regular enough, and persists long enough to allow it to reach thermal equilibrium with a specified thermometer, then it has a temperature according to that thermometer. An empirical thermometer registers degree of hotness for such a system. Such a temperature is called empirical. For example, Truesdell writes about classical thermodynamics: "At each time, the body is assigned a real number called the temperature. This number is a measure of how hot the body is. "

Physical systems that are too turbulent to have temperatures may still differ in hotness. A physical system that passes heat to another physical system is said to be the hotter of the two. More is required for the system to have a thermodynamic temperature. Its behavior must be so regular that its empirical temperature is the same for all suitably calibrated and scaled thermometers, and then its hotness is said to lie on the one-dimensional hotness manifold. This is part of the reason why heat is defined following Carathéodory and Born, solely as occurring other than by work or transfer of matter; temperature is advisedly and deliberately not mentioned in this now widely accepted definition.

This is also the reason why the zeroth law of thermodynamics is stated explicitly. If three physical systems, A, B, and C are each not in their own states of internal thermodynamic equilibrium, it is possible that, with suitable physical connections being made between them, A can heat B and B can heat C and C can heat A. In non-equilibrium situations, cycles of flow are possible. It is the special and uniquely distinguishing characteristic of internal thermodynamic equilibrium that this possibility is not open to thermodynamic systems (as distinguished amongst physical systems) which are in their own states of internal thermodynamic equilibrium; this is the reason why the zeroth law of thermodynamics needs explicit statement. That is to say, the relation 'is not colder than' between general non-equilibrium physical systems is not transitive, whereas, in contrast, the relation 'has no lower a temperature than' between thermodynamic systems in their own states of internal thermodynamic equilibrium is transitive. It follows from this that the relation 'is in thermal equilibrium with' is transitive, which is one way of stating the zeroth law.

Just as temperature may undefined for a sufficiently inhomogeneous system, so also may entropy be undefined for a system not in its own state of internal thermodynamic equilibrium. For example, 'the temperature of the solar system' is not a defined quantity. Likewise, 'the entropy of the solar system' is not defined in classical thermodynamics. It has not been possible to define non-equilibrium entropy, as a simple number for a whole system, in a clearly satisfactory way.

Rigorous definition of quantity of energy transferred as heat

It is sometimes convenient to have a rigorous definition of quantity of energy transferred as heat. Such a definition is customarily based on the work of Carathéodory (1909), referring to processes in a closed system, as follows.

The internal energy UX of a body in an arbitrary state X can be determined by amounts of work adiabatically performed by the body on its surrounds when it starts from a reference state O. Such work is assessed through quantities defined in the surroundings of the body. It is supposed that such work can be assessed accurately, without error due to friction in the surroundings; friction in the body is not excluded by this definition. The adiabatic performance of work is defined in terms of adiabatic walls, which allow transfer of energy as work, but no other transfer, of energy or matter. In particular they do not allow the passage of energy as heat. According to this definition, work performed adiabatically is in general accompanied by friction within the thermodynamic system or body. On the other hand, according to Carathéodory (1909), there also exist non-adiabatic walls, which are postulated to be "permeable only to heat", and are called diathermal.

For the definition of quantity of energy transferred as heat, it is customarily envisaged that an arbitrary state of interest Y is reached from state O by a process with two components, one adiabatic and the other not adiabatic. For convenience one may say that the adiabatic component was the sum of work done by the body through volume change through movement of the walls while the non-adiabatic wall was temporarily rendered adiabatic, and of isochoric adiabatic work. Then the non-adiabatic component is a process of energy transfer through the wall that passes only heat, newly made accessible for the purpose of this transfer, from the surroundings to the body. The change in internal energy to reach the state Y from the state O is the difference of the two amounts of energy transferred.

Although Carathéodory himself did not state such a definition, following his work it is customary in theoretical studies to define the quantity of energy transferred as heat, Q, to the body from its surroundings, in the combined process of change to state Y from the state O, as the change in internal energy, ÄUY, minus the amount of work, W, done by the body on its surrounds by the adiabatic process, so that Q = ÄUY " W.

In this definition, for the sake of conceptal rigour, the quantity of energy transferred as heat is not specified directly in terms of the non-adiabatic process. It is defined through knowledge of precisely two variables, the change of internal energy and the amount of adiabatic work done, for the combined process of change from the reference state O to the arbitrary state Y. It is important that this does not explicitly involve the amount of energy transferred in the non-adiabatic component of the combined process. It is assumed here that the amount of energy required to pass from state O to state Y, the change of internal energy, is known, independently of the combined process, by a determination through a purely adiabatic process, like that for the determination of the internal energy of state X above. The rigour that is prized in this definition is that there is one and only one kind of energy transfer admitted as fundamental: energy transferred as work. Energy transfer as heat is considered as a derived quantity. The uniqueness of work in this scheme is considered to guarantee rigor and purity of conception. The conceptual purity of this definition, based on the concept of energy transferred as work as an ideal notion, relies on the idea that some frictionless and otherwise non-dissipative processes of energy transfer can be realized in physical actuality. The second law of thermodynamics, on the other hand, assures us that such processes are not found in nature.

Heat, temperature, and thermal equilibrium regarded as jointly primitive notions

Before the rigorous mathematical definition of heat based on Carathéodory's 1909 paper, recounted just above, historically, heat, temperature, and thermal equilibrium were presented in thermodynamics textbooks as jointly primitive notions. Carathéodory introduced his 1909 paper thus: "The proposition that the discipline of thermodynamics can be justified without recourse to any hypothesis that cannot be verified experimentally must be regarded as one of the most noteworthy results of the research in thermodynamics that was accomplished during the last century. " Referring to the "point of view adopted by most authors who were active in the last fifty years", Carathéodory wrote: "There exists a physical quantity called heat that is not identical with the mechanical quantities (mass, force, pressure, etc.) and whose variations can be determined by calorimetric measurements. " James Serrin introduces an account of the theory of thermodynamics thus: "In the following section, we shall use the classical notions of heat, work, and hotnessas primitive elements,... That heat is an appropriate and natural primitive for thermodynamics was already accepted by Carnot. Its continued validity as a primitive element of thermodynamical structure is due to the fact that it synthesizes an essential physical concept, as well as to its successful use in recent work to unify different constitutive theories. " This traditional kind of presentation of the basis of thermodynamics includes ideas that may be summarized by the statement that heat transfer is purely due to spatial non-uniformity of temperature, and is by conduction and radiation, from hotter to colder bodies. It is sometimes proposed that this traditional kind of presentation necessarily rests on "circular reasoning"; against this proposal, there stands the rigorously logical mathematical development of the theory presented by Truesdell and Bharatha (1977).

This alternative approach to the definition of quantity of energy transferred as heat differs in logical structure from that of Carathéodory, recounted just above.

This alternative approach admits calorimetry as a primary or direct way to measure quantity of energy transferred as heat. It relies on temperature as one of its primitive concepts, and used in calorimetry. It is presupposed that enough processes exist physically to allow measurement of differences in internal energies. Such processes are not restricted to adiabatic transfers of energy as work. They include calorimetry, which is the commonest practical way of finding internal energy differences. The needed temperature can be either empirical or absolute thermodynamic.

In contrast, the Carathéodory way recounted just above does not use calorimetry or temperature in its primary definition of quantity of energy transferred as heat.

The Carathéodory way regards calorimetry only as a secondary or indirect way of measuring quantity of energy transferred as heat. As recounted in more detail just above, the Carathéodory way regards quantity of energy transferred as heat in a process as primarily or directly defined as a residual quantity. It is calculated from the difference of the internal energies of the initial and final states of the system, and from the actual work done by the system during the process. That internal energy difference is supposed to have been measured in advance through processes of purely adiabatic transfer of energy as work, processes that take the system between the initial and final states. By the Carathéodory way it is presupposed as known from experiment that there actually physically exist enough such adiabatic processes, so that there need be no recourse to calorimetry for measurement of quantity of energy transferred as heat. This presupposition is essential but is explicitly labeled neither as a law of thermodynamics nor as an axiom of the Carathéodory way. In fact, the actual physical existence of such adiabatic processes is indeed mostly supposition, and those supposed processes have in most cases not been actually verified empirically to exist.

Heat transfer in engineering

The discipline of heat transfer, typically considered an aspect of mechanical engineering and chemical engineering, deals with specific applied methods by which thermal energy in a system is generated, or converted, or transferred to another system. Although the definition of heat implicitly means the transfer of energy, the term heat transfer encompasses this traditional usage in many engineering disciplines and laymen language.

Heat transfer includes the mechanisms of heat conduction, thermal radiation, and mass transfer.

In engineering, the term convective heat transfer is used to describe the combined effects of conduction and fluid flow. From the thermodynamic point of view, heat flows into a fluid by diffusion to increase its energy, the fluid then transfers (advects) this increased internal energy (not heat) from one location to another, and this is then followed by a second thermal interaction which transfers heat to a second body or system, again by diffusion. This entire process is often regarded as an additional mechanism of heat transfer, although technically, "heat transfer" and thus heating and cooling occurs only on either end of such a conductive flow, but not as a result of flow. Thus, conduction can be said to "transfer" heat only as a net result of the process, but may not do so at every time within the complicated convective process.

Although distinct physical laws may describe the behavior of each of these methods, real systems often exhibit a complicated combination which are often described by a variety of complex mathematical methods.

7

CRYSTAL

Crystal, any solid material in which the component atoms are arranged in a definite pattern and whose surface regularity reflects its internal symmetry.

CLASSIFICATION

The definition of a solid appears obvious; a solid is generally thought of as being hard and firm. Upon inspection, however, the definition becomes less straightforward. A cube of butter, for example, is hard after being stored in a refrigerator and is clearly a solid. After remaining on the kitchen counter for a day, the same cube becomes quite soft, and it is unclear if the butter should still be considered a solid. Many crystals behave like butter in that they are hard at low temperatures but soft at higher temperatures. They are called solids at all temperatures below their melting point. A possible definition of a solid is an object that retains its shape if left undisturbed. The pertinent issue is how long the object keeps its shape. A highly viscous fluid retains its shape for an hour but not a year. A solid must keep its shape longer than that.

Basic Units of Solids

The basic units of solids are either atoms or atoms that have combined into molecules. The electrons of an atom move in orbits that form a shell structure around the nucleus. The shells are filled in a systematic order, with each shell accommodating only a small number of electrons. Different atoms have different numbers of electrons, which are distributed in a characteristic electronic structure of filled and partially filled shells. The arrangement of an atom's electrons determines its chemical properties. The properties of solids are usually predictable from the properties of their constituent atoms and molecules, and the different shell structures of atoms are therefore responsible for the diversity of solids.

All occupied shells of the argon (Ar) atom, for example, are filled, resulting in a spherical atomic shape. In solid argon the atoms are arranged according to the closest packing of these spheres.

The iron (Fe) atom, in contrast, has one electron shell that is only partially filled, giving the atom a net magnetic moment. Thus, crystalline iron is a magnet. The covalent bond between two carbon (C) atoms is the strongest bond found in nature. This strong bond is responsible for making diamond the hardest solid.

Long- and Short-range Order

A solid is crystalline if it has long-range order. Once the positions of an atom and its neighbours are known at one point, the place of each atom is known precisely throughout the crystal. Most liquids lack long-range order, although many have short-range order. Short range is defined as the first- or second-nearest neighbours of an atom. In many liquids the first-neighbour atoms are arranged in the same structure as in the corresponding solid phase.

At distances that are many atoms away, however, the positions of the atoms become uncorrelated. These fluids, such as water, have short-range order but lack long-range order. Certain liquids may have short-range order in one direction and long-range order in another direction; these special substances are called liquid crystals. Solid crystals have both short-range order and long-range order.

Solids that have short-range order but lack long-range order are called amorphous. Almost any material can be made amorphous by rapid solidification from the melt (molten state). This condition is unstable, and the solid will crystallize in time. If the timescale for crystallization is years, then the amorphous state appears stable. Glasses are an example of amorphous solids. In crystalline silicon (Si) each atom is tetrahedrally bonded to four neighbours. In amorphous silicon (a-Si) the same short-range order exists, but the bond directions become changed at distances farther away from any atom. Amorphous silicon is a type of glass. Quasicrystals are another type of solid that lack long-range order.

Most solid materials found in nature exist in polycrystalline form rather than as a single crystal. They are actually composed of millions of grains (small crystals) packed together to fill all space. Each individual grain has a different orientation than its neighbours. Although long-range order exists within one grain, at the boundary between grains, the ordering changes direction. A typical piece of iron or copper (Cu) is polycrystalline. Single crystals of metals are soft and malleable, while polycrystalline metals are harder and stronger and are more useful industrially.

Most polycrystalline materials can be made into large single crystals after extended heat treatment. In the past blacksmiths would heat a piece of metal to make it malleable: heat makes a few grains grow large by incorporating smaller ones. The smiths would bend the softened metal into shape and then pound it awhile; the pounding would make it polycrystalline again, increasing its strength.

Categories of Crystals

Crystals are classified in general categories, such as insulators, metals, semiconductors, and molecular solids. A single crystal of an insulator is usually transparent and resembles a piece of glass. Metals are shiny unless they have rusted. Semiconductors are sometimes shiny and sometimes transparent but are never rusty. Many crystals can be classified as a single type of solid, while others have intermediate behaviour.

Cadmium sulfide (CdS) can be prepared in pure form and is an excellent insulator; when impurities are added to cadmium sulfide, it becomes an interesting semiconductor. Bismuth (Bi) appears to be a metal, but the number of electrons available for electrical conduction is similar to that of semiconductors. In fact, bismuth is called a semimetal. Molecular solids are usually crystals formed from molecules or polymers. They can be insulating, semiconducting, or metallic, depending on the type of molecules in the crystal. New molecules are continuously being synthesized, and many are made into crystals. The number of different crystals is enormous.

Structure

Crystals can be grown under moderate conditions from all 92 naturally occurring elements except helium, and helium can be crystallized at low temperatures by using 25 atmospheres of pressure. Binary crystals are composed of two elements. There are thousands of binary crystals; some examples are sodium chloride (NaCl), alumina (Al2O3), and ice (H2O). Crystals can also be formed with three or more elements.

The Unit Cell

A basic concept in crystal structures is the unit cell. It is the smallest unit of volume that permits identical cells to be stacked together to fill all space. By repeating the pattern of the unit cell over and over in all directions, the entire crystal lattice can be constructed. A cube is the simplest example of a unit cell. Two other examples are shown in Figure. The first is the unit cell for a face-centred cubic lattice, and the second is for a body-centred cubic lattice.

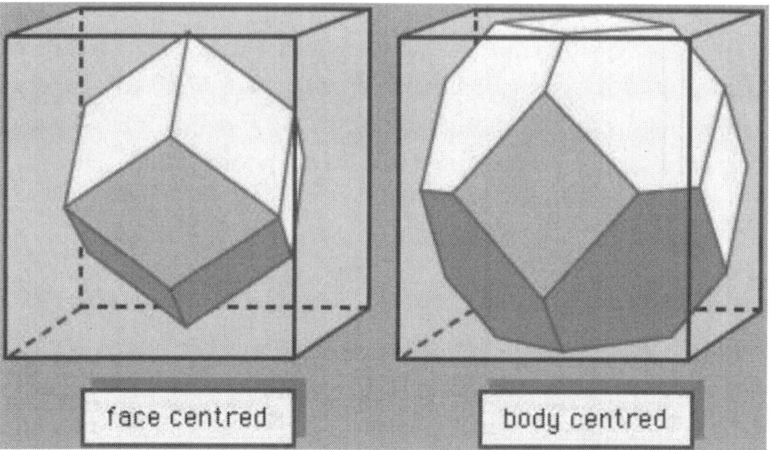

These structures are explained in the following paragraphs. There are only a few different unit-cell shapes, so many different crystals share a single unit-cell type. An important characteristic of a unit cell is the number of atoms it contains. The total number of atoms in the entire crystal is the number in each cell multiplied by the number of unit cells. Copper and aluminum (Al) each have one atom per unit cell, while zinc (Zn) and sodium chloride have two. Most crystals have only a few atoms per unit cell, but there are some exceptions. Crystals of polymers, for example, have thousands of atoms in each unit cell.

Structures of Metals

The elements are found in a variety of crystal packing arrangements. The most common lattice structures for metals are those obtained by stacking the atomic spheres into the most compact arrangement. There are two such possible periodic arrangements. In each, the first layer has the atoms packed into a plane-triangular lattice in which every atom has six immediate neighbours.

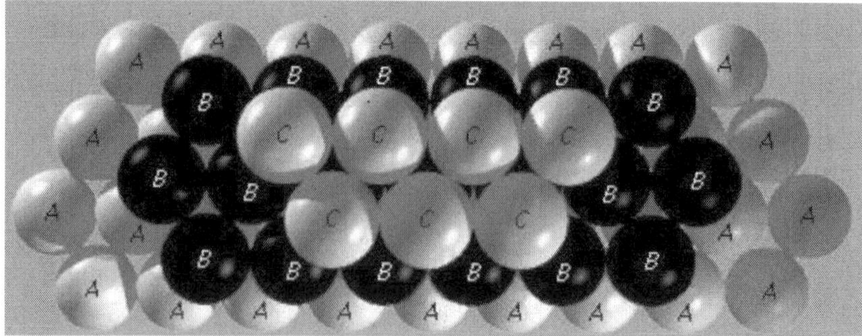

Figure shows this arrangement for the atoms labeled A. The second layer is shaded in the figure. It has the same plane-triangular structure; the atoms sit in the holes formed by the first layer. The first layer has two equivalent sets of holes, but the atoms of the second layer can occupy only one set. The third layer, labeled C, has the same structure, but there are two choices for selecting the holes that the atoms will occupy. The third layer can be placed over the atoms of the first layer, generating an alternate layer sequence ABABAB..., which is called the hexagonal- closest-packed (hcp) structure. Cadmium and zinc crystallize with this structure. The second possibility is to place the atoms of the third layer over those of neither of the first two but instead over the set of holes in the first layer that remains unoccupied. The fourth layer is placed over the first, and so there is a three-layer repetition ABCABCABC..., which is called the face-centred cubic (fcc), or cubic-closest-packed, lattice. Copper, silver (Ag), and gold (Au) crystallize in fcc lattices. In the hcp and the fcc structures the spheres fill 74 percent of the volume, which represents the closest possible packing of spheres. Each atom has 12 neighbours. The number of atoms in a unit cell is two for hcp structures and one for fcc. There are 32 metals that have the hcp lattice and 26 with the fcc. Another possible arrangement is the body-centred cubic (bcc) lattice, in which each atom has eight neighbours arranged at the corners of a cube. Figure below shows the cesium chloride (CsCl) structure, which is a cubic arrangement. If all atoms in this structure are of the same species, it is a bcc lattice. The spheres occupy 68 percent of the volume. There are 23 metals with the bcc arrangement. The sum of these three numbers (32 + 26 + 23) exceeds the number of elements that form metals (63), since some elements are found in two or three of these structures.

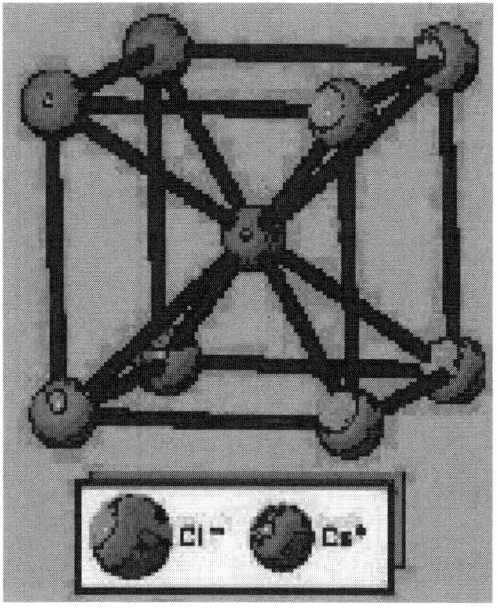

The fcc structure is also found for crystals of the rare gas solids neon (Ne), argon (Ar), krypton (Kr), and xenon (Xe). Their melting temperatures at atmospheric pressure are: Ne, 24.6 K; Ar, 83.8 K; Kr, 115.8 K; and Xe, 161.4 K.

Structures of Nonmetallic Elements

The elements in the fourth row of the periodic table—carbon, silicon, germanium (Ge), and á-tin (á-Sn)—prefer covalent bonding. Carbon has several possible crystal structures. Each atom in the covalent bond has four first-neighbours, which are at the corners of a tetrahedron.

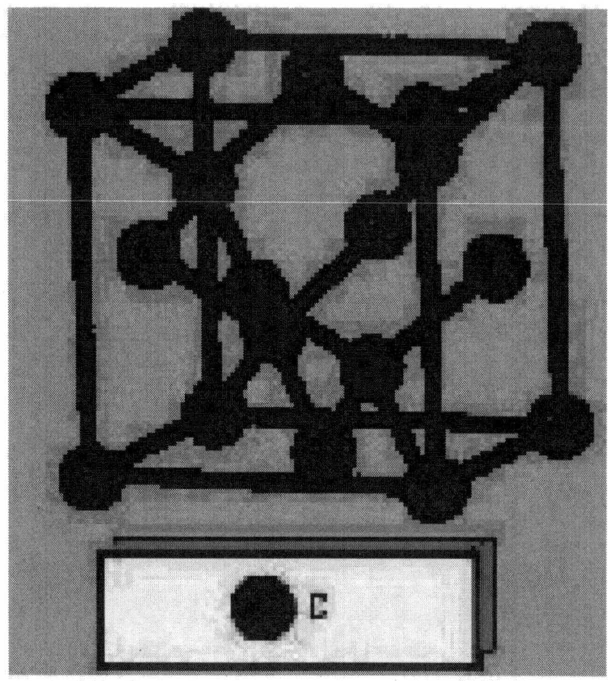

This arrangement is called the diamond lattice and is shown in Figure. There are two atoms in a unit cell, which is fcc. Large crystals of diamond are valuable gemstones. The crystal has other interesting properties; it has the highest sound velocity of any solid and is the best conductor of heat. Besides diamond, the other common form of carbon is graphite, which is a layered material. Each carbon atom has three coplanar near neighbours, forming an arrangement called the honeycomb lattice. Three-dimensional graphite crystals are obtained by stacking similar layers.

Another form of crystalline carbon is based on a molecule with 60 carbon atoms called buckminsterfullerene (C60). The molecular shape is spherical. Each carbon is bonded to three neighbours, as in graphite, and the spherical shape is achieved by a mixture of 12 rings with five sides and 20 rings with six sides. Similar structures were first visualized by the American architect R. Buckminster Fuller for geodesic domes. The C60 molecules, also called buckyballs, are quite strong and almost incompressible. Crystals are formed such that the balls are arranged in an fcc lattice with a one-nanometre spacing between the centres of adjacent balls. The similar C70 molecule has the shape of a rugby ball; C70 molecules also form an fcc crystal when stacked together. The solid fullerenes form molecular crystals, with weak binding—provided by van der Waals interactions—between the molecules.

Many elements form diatomic gases: hydrogen (H), oxygen (O), nitrogen (N), fluorine (F), chlorine (Cl), bromine (Br), and iodine (I). When cooled to low temperature, they form solids of diatomic molecules. Nitrogen has the hcp structure, while oxygen has a more complex structure.

The most interesting crystal structures are those of elements that are neither metallic, covalent, nor diatomic. Although boron (B) and sulfur (S) have several different crystal structures, each has one arrangement in which it is usually found. Twelve boron atoms form a molecule in the shape of an icosahedron (Figure 4). Crystals are formed by stacking the molecules. The â-rhombohedral structure of boron has seven of these icosahedral molecules in each unit cell, giving a total of 84 atoms. Molecules of sulfur are usually arranged in rings; the most common ring has eight atoms. The typical structure is á-sulfur, which has 16 molecules per unit cell, or 128 atoms. In the common crystals of selenium (Se) and tellurium (Te), the atoms are arranged in helical chains, which stack like cordwood. However, selenium also makes eight-atom rings, similar to sulfur, and forms crystals from them. Sulfur also makes helical chains, similar to selenium, and stacks them together into crystals.

Structures of Binary Crystals

Binary crystals are found in many structures. Some pairs of elements form more than one structure. At room temperature, cadmium sulfide may crystallize either in the zinc blende or wurtzite structure. Alumina also has two possible structures at room temperature, á-alumina (corundum) and â-alumina. Other binary crystals exhibit different structures at different temperatures. Among the most complex crystals are those of silicon dioxide (SiO2), which has seven different structures at various temperatures and pressures; the most common of these structures is quartz. Some pairs of elements form several different crystals in which the ions have different chemical valences. Cadmium (Cd) and phosphorus (P) form the crystals Cd3P2, CdP2, CdP4, Cd7P10, and Cd6P7. Only in the first case are the ions assigned the expected chemical valences of Cd2+ and P3-.

Among the binary crystals, the easiest structures to visualize are those with equal numbers of the two types of atoms. The structure of sodium chloride is based on a cube. To construct the lattice, the sodium and chlorine atoms are placed on alternate corners of a cube, and the structure is repeated. The structure of the sodium atoms alone, or the chlorine atoms alone, is fcc and defines the unit cell. The sodium chloride structure thus is made up of two interpenetrating fcc lattices. The cesium chloride lattice is based on the bcc structure; every other atom is cesium or chlorine. In this case, the unit cell is a cube.

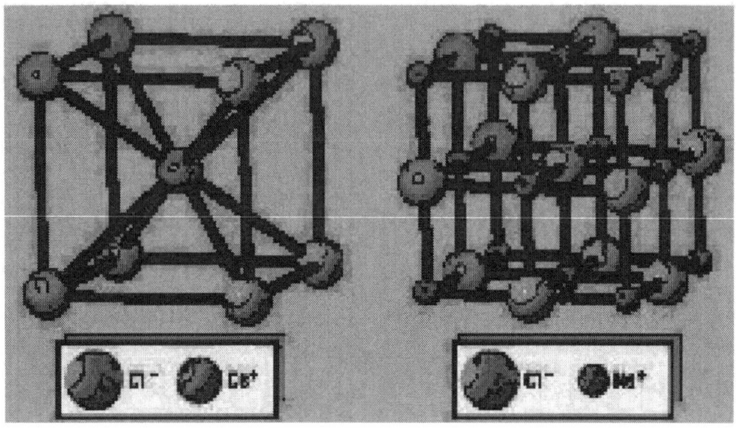

The third important structure for AB (binary) lattices is zinc blende.

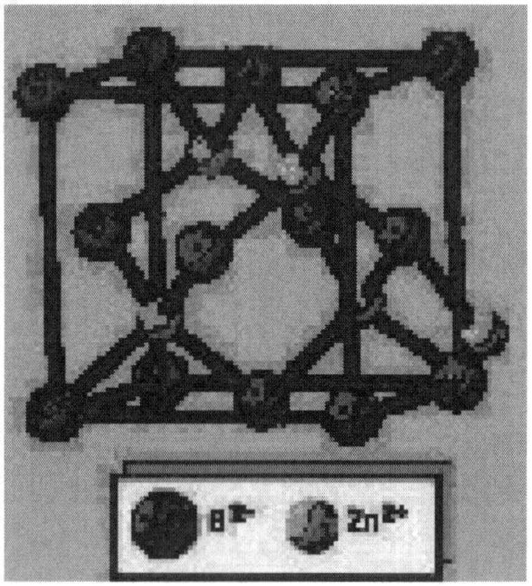

It is based on the diamond structure, where every other atom is A or B. Many binary semiconductors have this structure, including those with one atom from the third (boron, aluminum, gallium [Ga], or indium [In]) and one from the fifth (nitrogen, phosphorus, arsenic [As], or antimony [Sb]) column of the periodic table (GaAs, InP, etc.). Most of the chalcogenides (O, S, Se, Te) of cadmium and zinc (CdTe, ZnSe, ZnTe, etc.) also have the zinc blende structure. The mineral zinc blende is ZnS; its unit cell is also fcc. The wurtzite structure is based on the hcp lattice, where every other atom is A or B. These four structures comprise most of the binary crystals with equal numbers of cations and anions.

The fullerene molecule forms binary crystals MxC60 with alkali atoms, where M is potassium (K), rubidium (Rb), or cesium (Cs). The fullerene molecules retain their spherical shape, and the alkali atoms sit between them. The subscript x can take on several values. A compound with x = 6 (e.g., K6C60) is an insulator with the fullerenes in a bcc structure. The case x = 4 is an insulator with the body-centred tetragonal structure, while the case x = 3 is a metal with the fullerenes in an fcc structure. K3C60, Rb3C60, and Cs3C60 are superconductors at low temperatures.

Alloys

Alloys are solid mixtures of atoms with metallic properties. The definition includes both amorphous and crystalline solids. Although many pairs of elements will mix together as solids, many pairs will not. Almost all chemical entities can be mixed in liquid form. But cooling a liquid to form a solid often results in phase separation; a polycrystalline material is obtained in which each grain is purely one atom or the other. Extremely rapid cooling can produce an amorphous alloy. Some pairs of elements form alloys that are metallic crystals. They have useful properties that differ from those exhibited by the pure elements. For example, alloying makes a metal stronger; for this reason alloys of gold, rather than the pure metal, are used in jewelry.

Atoms tend to form crystalline alloys when they are of similar size. The sizes of atoms are not easy to define, however, because atoms are not rigid objects with sharp boundaries. The outer part of an atom is composed of electrons in bound orbits; the average number of electrons decreases gradually with increasing distance from the nucleus. There is no point that can be assigned as the precise radius of the atom. Scientists have discovered, however, that each atom in a solid has a characteristic radius that determines its preferred separation from neighbouring atoms. For most types of atom this radius is constant, even in different solids. An empirical radius is assigned to each atom for bonding considerations, which leads to the concept of atomic size. Atoms readily make crystalline alloys when the radii of the two types of atoms agree to within roughly 15 percent.

Two kinds of ordering are found in crystalline alloys. Most alloys at low temperature are binary crystals with perfect ordering. An example is the alloy of copper and zinc. Copper is fcc, whereas zinc is hcp. A 50-percent-zinc–50-percent-copper alloy has a different structure—â-brass. At low temperatures it has the cesium chloride structure: a bcc lattice with alternating atoms of copper and zinc and a cubic unit cell. If the temperature is raised above 470° C, however, a phase transition to another crystalline state occurs. The ordering at high temperature is also bcc, but now each site has equal probability of having a copper or zinc atom. The two types of atoms randomly occupy each site, but there is still long-range order. At all temperatures, thousands of atoms away from a site, the location of the atom site can be predicted with certainty. At temperatures below 470° C one also knows whether that site will be occupied by a copper or zinc atom, while above 470° C there is an equal likelihood of finding either atom. The high-temperature phase is crystalline but disordered. The disorder phase is obtained through a partial melting, not into a liquid state but into a less ordered one. This behaviour is typical of metal alloys. Other common alloys are steel, an alloy of iron and carbon; stainless steel, an alloy of iron, nickel (Ni), and chromium (Cr); pewter and solder, alloys of tin and lead (Pb); and britannia metal, an alloy of tin, antimony, and copper.

CRYSTAL DEFECTS

A crystal is never perfect; a variety of imperfections can mar the ordering. A defect is a small imperfection affecting a few atoms. The simplest type of defect is a missing atom and is called a vacancy. Since all atoms occupy space, extra atoms cannot be located at the lattice sites of other atoms, but they can be found between them; such atoms are called interstitials. Thermal vibrations may cause an atom to leave its original crystal site and move into a nearby interstitial site, creating a vacancy-interstitial pair. Vacancies and interstitials are the types of defects found in a pure crystal. In another defect, called an impurity, an atom is present that is different from the host crystal atoms. Impurities may either occupy interstitial spaces or substitute for a host atom in its lattice site.

There is no sharp distinction between an alloy and a crystal with many impurities. An alloy results when a sufficient number of impurities are added that are soluble in the host metal. However, most elements are not soluble in most crystals. Crystals generally can tolerate a few impurities per million host atoms. If too many impurities of the insoluble variety are added, they coalesce to form their own small crystallite. These inclusions are called precipitates and constitute a large defect.

Germanium is a common impurity in silicon. It prefers the same tetrahedral bonding as silicon and readily substitutes for silicon atoms. Similarly, silicon is a common impurity in germanium. No large crystal can be made without impurities;

the purest large crystal ever grown was made of germanium. It had about 1010 impurities in each cubic centimetre of material, which is less than one impurity for each trillion atoms.

Impurities often make crystals more useful. In the absence of impurities, á-alumina is colourless. Iron and titanium impurities impart to it a blue colour, and the resulting gem-quality mineral is known as sapphire. Chromium impurities are responsible for the red colour characteristic of rubies, the other gem of á-alumina. Pure semiconductors rarely conduct electricity well at room temperatures. Their ability to conduct electricity is caused by impurities. Such impurities are deliberately added to silicon in the manufacture of integrated circuits. In fluorescent lamps the visible light is emitted by impurities in the phosphors (luminescent materials).

Other imperfections in crystals involve many atoms. Twinning is a special type of grain boundary defect, in which a crystal is joined to its mirror image. Another kind of imperfection is a dislocation, which is a line defect that may run the length of the crystal. One of the many types of dislocations is due to an extra plane of atoms that is inserted somewhere in the crystal structure.

Another type, called an edge dislocation, is shown in Figure. This line defect occurs when there is a missing row of atoms. In the figure the crystal arrangement is perfect on the top and on the bottom. The defect is the row of atoms missing from region b. This mistake runs in a line that is perpendicular to the page and places a strain on region a.

Dislocations are formed when a crystal is grown, and great care must be taken to produce a crystal free of them. Dislocations are stable and will exist for years. They relieve mechanical stress. If one presses on a crystal, it will accommodate the induced stress by growing dislocations at the surface, which gradually move inward. Dislocations make a crystal mechanically harder. When a metal bar is cold-worked by rolling or hammering, dislocations and grain boundaries are introduced; this causes the hardening.

Determination of Crystal Structures

Crystal structures are determined by scattering experiments using a portion of the crystal as the target. A beam of particles is sent toward the target, and upon impact some of the particles scatter from the crystal and ricochet in various directions. A measurement of the scattered particles provides raw data, which is then computer-processed to give a picture of the atomic arrangements. The positions are then inferred from the computer-analyzed data.

Max von Laue first suggested in 1912 that this measurement could be done using X rays, which are electromagnetic radiation of very high frequency. High frequencies are needed because these waves have a short wavelength. Von Laue realized that atoms have a spacing of only a few angstroms (1 angstrom [Å] is

10–10 metre, or 3.94 × 10–9 inch). In order to measure atomic arrangements, the particles scattering from the target must also have a wavelength of a few angstroms. X rays are required when the beam consists of electromagnetic radiation. The X rays only scatter in certain directions, and there are many X rays associated with each direction. The scattered particles appear in spots corresponding to locations where the scattering from each identical atom produces an outgoing wave that has all the wavelengths in phase. Figure 6 shows incoming waves in phase. The scattering from atom A2 has a longer path than that from atom A1. If this additional path has a length (AB + BC) that is an exact multiple of the wavelength, then the two outgoing waves are in phase and reinforce each other. If the scattering angle is changed slightly, the waves no longer add coherently and begin to cancel one another. Combining the scattered radiation from all the atoms in the crystal causes all the outgoing waves to add coherently in certain directions and produce a strong signal in the scattered wave. If the extra path length (AB + BC) is five wavelengths, for example, the spot appears in one place. If it is six wavelengths, the spot is elsewhere. Thus, the different spots correspond to the different multiples of the wavelength of the X ray. The measurement produces two types of information: the directions of the spots and their intensity. This information is insufficient to deduce the exact crystal structure, however, as there is no algorithm by which the computer can go directly from the data to the structure. The crystallographer must propose various structures and compute how they would scatter the X rays. The theoretical results are compared with the measured one, and the theoretical arrangement is chosen that best fits the data. Although this procedure is fast when there are only a few atoms in a unit cell, it may take months or years for complex structures. Some protein molecules, for instance, have hundreds of atoms. Crystals of the proteins are grown, and X rays are used to measure the structure. The goal is to determine how the atoms are arranged in the protein, rather than how the proteins are arranged in the crystal.

Beams of neutrons may also be used to measure crystal structure. The beam of neutrons is obtained by drilling a hole in the side of a nuclear reactor. The energetic neutrons created in nuclear fission escape through the hole. The motion of elementary particles is governed by quantum, or wave, mechanics. Each neutron has a wavelength that depends on its momentum. The scattering directions are determined by the wavelength, as is the case with X rays. The wavelengths for neutrons from a reactor are suitable for measuring crystal structures.

X rays and neutrons provide the basis for two competing technologies in crystallography. Although they are similar in principle, the two methods have some differences. X rays scatter from the electrons in the atoms so that more electrons result in more scattering. X rays easily detect atoms of high atomic number, which have many electrons, but cannot readily locate atoms with few electrons. In hydrogen-bonded crystals, X rays do not detect the protons at all. Neutrons, on the other hand, scatter from the atomic nucleus. They scatter readily from protons and are excellent for determining the structure of hydrogen-bonded solids. One drawback to this method is that some nuclei absorb neutrons completely, and there is little scattering from these targets.

Beams of electrons can also be used to measure crystal structure, because energetic electrons have a wavelength that is suitable for such measurements. The problem with electrons is that they scatter strongly from atoms. Proper interpretation of the experimental results requires that an electron scatter only from one atom and leave the crystal without scattering again. Low-energy electrons scatter many times, and the interpretation must reflect this. Low-energy electron diffraction (LEED) is a technique in which a beam of electrons is directed toward the surface. The scattered electrons that reflect backward from the surface are measured. They scatter many times before leaving backward but mainly leave in a few directions that appear as "spots" in the measurements. An analysis of the varied spots gives information on the crystalline arrangement. Because the electrons are scattered strongly by the atoms in the first few layers of the surface, the measurement gives only the arrangements of atoms in these layers. It is assumed that the same structure is repeated throughout the crystal. Another scattering experiment involves electrons of extremely high energy. The scattering rate decreases as the energy of the electron increases, so that very energetic electrons usually scatter only once. Various electron microscopes are constructed on this principle.

Types of Bonds

The properties of a solid can usually be predicted from the valence and bonding preferences of its constituent atoms. Four main bonding types are discussed here: ionic, covalent, metallic, and molecular. Hydrogen-bonded solids, such as ice, make up another category that is important in a few crystals. There are many examples of solids that have a single bonding type, while other solids have a mixture of types, such as covalent and metallic or covalent and ionic.

Ionic Bonds

Sodium chloride exhibits ionic bonding. The sodium atom has a single electron in its outermost shell, while chlorine needs one electron to fill its outer shell. Sodium donates one electron to chlorine, forming a sodium ion (Na+) and a chlorine ion (Cl"). Each ion thus attains a closed outer shell of electrons and takes on a spherical shape. In addition to having filled shells and a spherical shape, the ions of an ionic solid have integer valence. An ion with positive valence is called a cation. In an ionic solid the cations are surrounded by ions with negative valence, called anions. Similarly, each anion is surrounded by cations. Since opposite charges attract, the preferred bonding occurs when each ion has as many neighbours as possible, consistent with the ion radii. Six or eight nearest neighbours are typical; the number depends on the size of the ions and not on the bond angles. The alkali halide crystals are binaries of the AH type, where A is an alkali ion (lithium [Li], sodium, potassium, rubidium, or cesium) and H is a halide ion (fluorine, chlorine, bromine, or iodine). The crystals have ionic bonding, and each ion has six or eight neighbours. Metal ions in the alkaline earth series (magnesium [Mg], calcium [Ca], barium [Ba], and strontium [Sr]) have two electrons in their outer shells and form divalent cations in ionic crystals. The chalcogenides (oxygen, sulfur, selenium, and tellurium) need two electrons to fill their outer p-shell. (Electron shells are divided into subshells, designated as s, p, d, f, g, and so forth. Each subshell is divided further into orbitals.) Two electrons are transferred from the cations to the anions, leaving each with a closed shell. The alkaline earth chalcogenides form ionic binary crystals such as barium oxide (BaO), calcium sulfide (CaS), barium selenide (BaSe), or strontium oxide (SrO). They have the same structure as sodium chloride, with each atom having six neighbours. Oxygen can be combined with various cations to form a large number of ionically bonded solids.

Covalent Bonds

Silicon, carbon, germanium, and a few other elements form covalently bonded solids. In these elements there are four electrons in the outer sp-shell, which is half filled. (The sp-shell is a hybrid formed from one s and one p subshell.) In the covalent bond an atom shares one valence (outer-shell) electron with each of its four nearest neighbour atoms. The bonds are highly directional and prefer a tetrahedral arrangement. A covalent bond is formed by two electrons—one from each atom—located in orbitals between the ions. Insulators, in contrast, have all their electrons within shells inside the atoms.

The perpetual spin of an electron is an important aspect of the covalent bond. From a vantage point above the spinning particle, counterclockwise rotation is designated spin-up, while clockwise rotation is spin-down. A fundamental law of quantum physics is the Pauli exclusion principle, which states that no two electrons can occupy the same point in space at the same time with the same direction of spin. In a covalent bond two electrons occupy the same small volume of space (i.e., the same orbital) at all times, so they must have opposite spin: one up and one down. The exclusion principle is then satisfied, and the resulting bond is strong.

In graphite the carbon atoms are arranged in parallel sheets, and each atom has only three near neighbours. The covalent bonds between adjacent carbons within each layer are quite strong and are called ó bonds. The fourth valence electron in carbon has its orbital perpendicular to the plane. This orbital bonds weakly with the similar orbitals on all three neighbours, forming ð bonds. The four bonds for each carbon atom in the graphite structure are not arranged in a tetrahedron; three are in a plane. The planar arrangement results in strong bonding, although not as strong as the bonding in the diamond configuration. The bonding between layers is quite weak and arises from the van der Waals interaction; there is much slippage parallel to the layers. Diamond and graphite form an interesting contrast: diamond is the hardest material in nature and is used as an abrasive, while graphite is used as a lubricant.

Besides the elemental semiconductors, such as silicon and germanium, some binary crystals are covalently bonded. Gallium has three electrons in the outer shell, while arsenic lacks three. Gallium arsenide (GaAs) could be formed as an insulator by transferring three electrons from gallium to arsenic; however, this does not occur. Instead, the bonding is more covalent, and gallium arsenide is a covalent semiconductor. The outer shells of the gallium atoms contribute three electrons, and those of the arsenic atoms contribute five, providing the eight electrons needed for four covalent bonds. The centres of the bonds are not at the midpoint between the ions but are shifted slightly toward the arsenic. Such bonding is typical of the III–V semiconductors—i.e., those consisting of one element from the third column of the periodic table and one from the fifth column. Elements from the third column (boron, aluminum, gallium, and indium) contribute three electrons, while the fifth-column elements (nitrogen, phosphorus, arsenic, and antimony) contribute five electrons. All III–V semiconductors are covalently bonded and typically have the zinc blende structure with four neighbours per atom. Most common semiconductors favour this arrangement.

The factor that determines whether a binary crystal will act as an insulator or a semiconductor is the valence of its constituent atoms. Ions that donate or accept one or two valence electrons form insulators. Those that have three to five valence electrons tend to have covalent bonds and form semiconductors. There are exceptions to these rules, however, as is the case with the IV–VI semiconductors such as lead sulfide. Heavier elements from the fourth column of the periodic table (germanium, tin, and lead) combine with the chalcogenides from the sixth row to form good binary semiconductors such as germanium telluride (GeTe) or tin sulfide (SnS). They have the sodium chloride structure, where each atom has six neighbours. Although not tetrahedrally bonded, they are good semiconductors.

Filled atomic shells with d-orbitals have an important role in covalent bonding. Electrons in atomic orbits have angular momentum (L), which is quantized in integer (n) multiples of Planck's constant h: $L = nh$. Electron orbitals with $n = 0$ are called s-states, with $n = 1$ are p-states, and with $n = 2$ are d-states. Silver and copper ions have one valence electron outside their closed shells. The outermost filled shell is a d-state and affects the bonding. Eight binary crystals are formed from the copper and silver halides. Three (AgF, AgCl, AgBr) have the sodium chloride structure with six neighbours. The other five (AgI, CuF, CuCl, CuBr, CuI) have the zinc blende structure with four neighbours. The bonding in this group of solids is on the borderline between covalent and ionic, since the crystals prefer both types of bonds. The alkali metal halides exhibit somewhat different behaviour. The alkali metals are also monovalent cations, but their halides are strictly ionic. The difference in bonding between the alkali metals on the one hand and silver and copper on the other hand is that silver and copper have filled d-shells while the alkalis have filled p-shells. Since the d-shells are filled, they do not covalently bond. This group of electrons is, however, highly polarizable, which influences the bonding of the valence electrons. Similar behaviour is found for zinc and cadmium, which have two valence electrons outside a filled d-shell. They form binary crystals with the chalcogenides, which have tetrahedral bonding. In this case the covalent bonding seems to be preferred over the ionic bond. In contrast, the alkaline earth chalcogenides, which are also divalent, have outer p-shells and are ionic. The zinc and cadmium chalcogenides are covalent, as the outer d-shell electrons of the two cations favour covalent bonding.

Metallic Bonds

Metallic bonds fall into two categories. The first is the case in which the valence electrons are from the sp-shells of the metal ions; this bonding is quite weak. In the second category the valence electrons are from partially filled d-shells, and this bonding is quite strong. The d-bonds dominate when both types of bonding are present.

The simple metals are bonded with sp-electrons. The electrons of these metal atoms are in filled atomic shells except for a few electrons that are in unfilled sp-shells. The electrons from the unfilled shells are detached from the metal ion and are free to wander throughout the crystal. They are called conduction electrons, since they are responsible for the electrical conductivity of metals. Although the conduction electrons may roam anywhere in the crystal, they are distributed uniformly throughout the entire solid. Any large imbalance of charge is prevented by the strong electrical attraction between the negative electrons and the positive ions, plus the strong repulsion between electrons. The phrase electron correlation describes the correlated movements of the electrons; the motion of each electron depends on the positions of neighbouring electrons. Electrons have strong short-range order with one another. Correlation ensures that each unit cell in the crystal has, on the average, the number of electrons needed to cancel the positive charge of the cation so that the unit cell is electrically neutral.

Cohesive energy is the energy gained by arranging the atoms in a crystalline state, as compared with the gas state. Insulators and semiconductors have large cohesive energies; these solids are bound together strongly and have good mechanical strength. Metals with electrons in sp-bonds have very small cohesive energies. This type of metallic bond is weak; the crystals are barely held together. Single crystals of simple metals such as sodium are mechanically weak. At room temperature the crystals have the mechanical consistency of warm butter. Special care must be used in handling these crystals, because they are easily distorted. Metals such as magnesium or aluminum must be alloyed or polycrystalline to have any mechanical strength. Although the simple metals are found in a variety of structures, most are in one of the three closest-packed structures: fcc, bcc, and hcp. Theoretical calculations show that the cohesive energy of a given metal is almost the same in each of the different crystal arrangements; therefore, crystal arrangements are unimportant in metals bound with electrons from sp-shells.

A different type of metallic bonding is found in transition metals, which are metals whose atoms are characterized by unfilled d-shells. The d-orbitals are more tightly bound to an ion than the sp-orbitals. Electrons in d-shells do not wander away from the ion. The d-orbitals form a covalent bond with the d-orbitals on the neighbouring atoms. The bonding of d-orbitals does not occur in a tetrahedral arrangement but has a different directional preference. In metals the bonds from d-orbitals are not completely filled with electrons. This situation is different from the tetrahedral bonds in semiconductors, which are filled with eight electrons. In transition metals the covalent bonds formed with the d-electrons are much stronger than the weak bonds made with the sp-electrons of simple metals. The cohesive energy is much larger in transition metals. Titanium, iron, and tungsten, for example, have exceptional mechanical strength. Crystal arrangements are important in the behaviour of the transition metals and occur in the close-packed fcc, bcc, or hcp arrangements.

MOLECULAR BINDING

The Dutch physicist Johannes D. van der Waals first proposed the force that binds molecular solids. Any two atoms or molecules have a force of attraction (F) that varies according to the inverse seventh power of the distance R between the centres of the atoms or molecules: F = "C/R7, where C is a constant. The force, known as the van der Waals force, declines rapidly with the distance R and is quite weak. If the atoms or molecules have a net charge, there is a strong force whose strength varies according to Coulomb's law as the inverse second power of the separation distance: F = –C2 /R2, where C2 is a constant. This force provides the binding in ionic crystals and some of the binding in metals. Coulomb's law does not apply to atoms or molecules without a net charge. Molecules with a dipole moment, such as water, have a strong attractive force owing to the interactions between the dipoles. For atoms and molecules with neither net charges nor dipole moments, the van der Waals force provides the crystal binding. The force of gravity also acts between neutral atoms and molecules, but it is far too weak to bind molecules into crystals.

The van der Waals force is caused by quantum fluctuations. Two neighbouring atoms that are each fluctuating can lower their joint energy by correlating their fluctuations. The van der Waals force arises from correlations in their dipole fluctuations. Electrons bound in atomic orbits are in constant motion around the nucleus, and the distribution of charges in the atom changes constantly as the electrons move, owing to quantum fluctuations. One fluctuation might produce a momentary electric dipole moment (i.e., a separation of charges) on an atom if a majority of its electrons are on one side of the nucleus. The dipole moment creates an electric field on a neighbouring atom; this field will induce a dipole moment on the second atom. The two dipoles attract one another via the van der Waals interaction. Since the force depends on the inverse seventh power of the distance, it declines rapidly with increasing distance. Atoms have a typical radius of one to three angstroms. The van der Waals force binds atoms and molecules within a few angstroms of each other; beyond that distance the force is negligible. Although weak, the van der Waals force is always present and is important in cases where the other forces are absent.

Hydrogen is rarely found as a single atom. Instead it forms diatomic molecules (H2), which are gaseous at room temperature. At lower temperatures the hydrogen becomes a liquid and at about 20 K turns into a solid. The molecule retains its identity in the liquid and solid states. The solid exists as a molecular crystal of covalently bound H2 molecules. The molecules attract one another by van der Waals forces, which provide the crystal binding. Helium, the second element in the periodic table, has two electrons, which constitute a filled atomic shell. In its liquid and solid states, the helium atoms are bound together by van der Waals forces. In fact, all the rare gases (helium, neon, argon, krypton, and xenon) are molecular crystals with the binding provided by van der Waals forces.

Many organic molecules form crystals where the molecules are bound by van der Waals forces. In methane (CH4), a central carbon makes a covalent bond with each hydrogen atom, forming a tetrahedron. In crystalline methane the molecules are arranged in the fcc structure. Benzene (C6H6) has the carbon atoms in a hexagonal ring; each carbon has three coplanar ó bonds, as in graphite, where two bonds are with neighbouring carbon atoms and the third bond is with a hydrogen atom. Crystalline benzene has four molecules per unit cell in a complex arrangement. Fullerene and the rare gas atoms are spherical, and the crystalline arrangement corresponds to the closest packing of spheres. Most organic molecules, however, are not spherical and display irregular shapes. For odd-shaped molecules, the van der Waals interaction depends on the rotational orientation of the two molecules. In order to maximize the force, the molecules in the crystal have unusual arrangements, as in the case of benzene.

HYDROGEN BONDING

Hydrogen bonding is important in a few crystals, notably in ice. With its lone electron, a hydrogen atom usually forms a single covalent bond with an electronegative atom. In the hydrogen bond the atom is ionized to a proton. The proton sits between two anions and joins them. Hydrogen bonding occurs with only the most electronegative ions: nitrogen, oxygen, and fluorine. In water the hydrogen links pairs of oxygen ions. Water is found in many different crystal structures, but they all have the feature that the hydrogen atoms sit between pairs of oxygen. Another hydrogen-bonded solid is hydrogen fluoride (HF), in which the hydrogen atom (proton) links pairs of fluorines.

Crystal Growth

The earliest crystal grower was nature. Many excellent crystals of minerals formed in the geologic past are found in mines and caves throughout the world. Most precious and semiprecious stones are well-formed crystals. Early efforts to produce synthetic crystals were concentrated on making gems. Synthetic ruby was grown by the French scientist Marc Antoine Augustin Gaudin in 1873. Since about 1950 scientists have learned to grow in the laboratory crystals of quality equal or superior to those found in nature. New techniques for growth are continually being developed, and crystals with three or more atoms per unit cell are continually being discovered.

Vapour Growth

Crystals can be grown from a vapour when the molecules of the gas attach themselves to a surface and move into the crystal arrangement. Several important conditions must be met for this to occur. At constant temperature and equilibrium conditions, the average number of molecules in the gas and solid states is constant; molecules leave the gas and attach to the surface at the same rate that they leave the surface to become gas molecules. For crystals to grow, the gas-solid chemical system must be in a nonequilibrium state such that there are too many gaseous molecules for the conditions of pressure and temperature. This state is called supersaturation. Molecules are more prone to leave the gas than to rejoin it, so they become deposited on the surface of the container. Supersaturation can be induced by maintaining the crystal at a lower temperature than the gas. A critical stage in the growth of a crystal is seeding, in which a small piece of crystal of the proper structure and orientation, called a seed, is introduced into the container. The gas molecules find the seed a more favourable surface than the walls and preferentially deposit there. Once the molecule is on the surface of the seed, it wanders around this surface to find the preferred site for attachment. Growth proceeds one molecule at a time and one layer at a time. The process is slow; it takes days to grow a small crystal. Crystals are grown at temperatures well below the melting point to reduce the density of defects. The advantage of vapour growth is that very pure crystals can be grown by this method, while the disadvantage is that it is slow.

Most clouds in the atmosphere are ice crystals that form by vapour growth from water molecules. Most raindrops are crystals as they begin descending but thaw during their fall to Earth. Seeding for rain—accomplished by dropping silver iodide crystals from airplanes—is known to induce precipitation. In the laboratory, vapour growth is usually accomplished by flowing a supersaturated gas over a seed crystal. Quite often a chemical reaction at the surface is needed to deposit the atoms. Crystals of silicon can be grown by flowing chlorosilane (SiCl4) and hydrogen (H2) over a seed crystal of silicon. Hydrogen acts as the buffer gas by controlling the temperature and rate of flow. The molecules dissociate on the surface in a chemical reaction that forms hydrogen chloride (HCl) molecules. Hydrogen chloride molecules leave the surface, while silicon atoms remain to grow into a crystal.

Binary crystals such as gallium arsenide (GaAs) are grown by a similar method. One process employs gallium chloride (GaCl) as the gallium carrier. Arsenic is provided by molecules such as arsenous chloride (AsCl3), arsine (AsH3), or As4 (yellow arsenic). These molecules, with hydrogen as the buffer gas, grow crystals of gallium arsenide while forming gas molecules such as gallium trichloride (GaCl3) and hydrogen chloride. Trimethylgallium, (CH3)3Ga, is another molecule that can be used to deliver gallium to the surface.

GROWTH FROM SOLUTION

Large single crystals may be grown from solution. In this technique the seed crystal is immersed in a solvent that contains typically about 10–30 percent of the desired solute. The choice of solvent usually depends on the solubility of the solute. The temperature and pH (a measure of acidity or basicity) of the solution must be well controlled. The method is faster than vapour growth, because there is a higher concentration of molecules at the surface in a liquid as compared to a gas, but it is still relatively slow.

Growth from the Melt

This method is the most basic. A gas is cooled until it becomes a liquid, which is then cooled further until it becomes a solid. Polycrystalline solids are typically produced by this method unless special techniques are employed. In any case, the temperature must be controlled carefully. Large crystals can be grown rapidly from the liquid elements using a popular method invented in 1918 by the Polish scientist Jan Czochralski and called crystal pulling. One attaches a seed crystal to the bottom of a vertical arm such that the seed is barely in contact with the material at the surface of the melt.

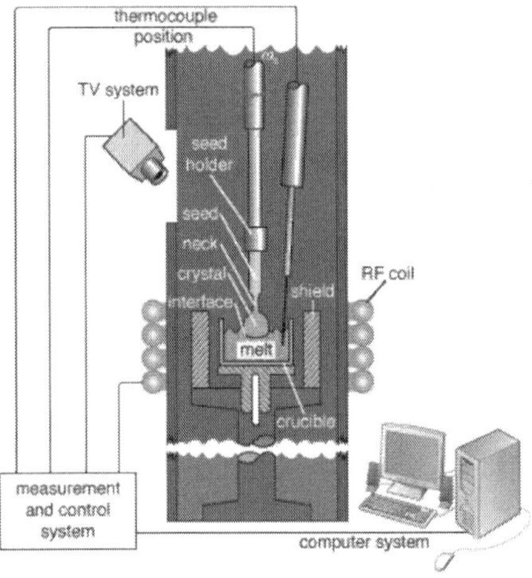

A modern Czochralski apparatus is shown in Figure. The arm is raised slowly, and a crystal grows underneath at the interface between the crystal and the melt. Usually the crystal is rotated slowly, so that inhomogeneities in the liquid are not replicated in the crystal. Large-diameter crystals of silicon are grown in this way for use as computer chips. Based on measurements of the weight of the crystal during the pulling process, computer-controlled apparatuses can vary the pulling rate to produce any desired diameter. Crystal pulling is the least expensive way to grow large amounts of pure crystal. The original seed is on the right tip. Binary crystals can also be pulled; for example, synthetic sapphire crystals can be pulled from molten alumina. Special care is required to grow binary and other multicomponent crystals; the temperature must be precisely controlled because such crystals may be grown only at a single, extremely high temperature. The melt has a tendency to be inhomogeneous, since the two liquids may try to separate by gravity.

The Bridgman method (named after the American scientist Percy Williams Bridgman) is also widely used for growing large single crystals. The molten material is put into a crucible, often of silica, which has a cylindrical shape with a conical lower end. Heaters maintain the molten state. As the crucible is slowly lowered into a cooler region, a crystal starts growing in the conical tip. The crucible is lowered at a rate that matches the growth of the crystal, so that the interface between crystal and melt is always at the same temperature. The rate of moving the crucible depends on the temperature and the material. When done successfully, the entire molten material in the crucible grows into a single large crystal. One disadvantage of the method is that excess impurities are pushed out of the crystal during growth. A layer of impurities grows at the interface between melt and solid as this surface moves up the melt, and the impurities become concentrated in the higher part of the crystal.

Epitaxy is the technique of growing a crystal, layer by layer, on the atomically flat surface of another crystal. In homoepitaxy a crystal is grown on a substrate of the same material. Silicon layers of different impurity content, for example, are grown on silicon substrates in the manufacture of computer chips. Heteroepitaxy, on the other hand, is the growth of one crystal on the substrate of another. Silicon substrates are often used since they are readily available in atomically smooth form. Many different semiconductor crystals can be grown on silicon, such as gallium arsenide, germanium, cadmium telluride (CdTe), and lead telluride (PbTe). Any flat substrate can be used for epitaxy, however, and insulators such as rock salt (NaCl) and magnesium oxide (MgO) are also used.

Molecular-beam epitaxy, commonly abbreviated as MBE, is a form of vapour growth. The field began when the American scientist John Read Arthur reported in 1968 that gallium arsenide could be grown by sending a beam of gallium atoms and arsenic molecules toward the flat surface of a crystal of the molecule. The amount of gas molecules can be controlled to grow just one layer, or just two, or any desired amount. This method is slow, since molecular beams have low densities of atoms. Chemical vapour deposition (CVD) is another form of epitaxy that makes use of the vapour growth technique. Also known as vapour-phase epitaxy (VPE), it is much faster than MBE since the atoms are delivered in a flowing gas rather than in a molecular beam. Synthetic diamonds are grown by CVD. Rapid growth occurs when methane (CH4) is mixed with atomic hydrogen gas, which serves as a catalyst. Methane dissociates on a heated surface of diamond. The carbon remains on the surface, and the hydrogen leaves as a molecule. Growth rates are several micrometres (1 micrometre is equal to 0.00004 inch) per hour. At that rate, a stone 1 centimetre (0.4 inch) thick is grown in 18 weeks. CVD diamonds are of poor quality as gemstones but are important electronic materials. Because hydrogen is found in nature as a molecule rather than as a single atom, making atomic hydrogen gas is the major expense in growing CVD diamonds. Liquid-phase epitaxy (LPE) uses the solution method to grow crystals on a substrate. The substrate is placed in a solution with a saturated concentration of solute. This technique is used to grow many crystals employed in modern electronics and optoelectronic devices, such as gallium arsenide, gallium aluminum arsenide, and gallium phosphide.

An important concern in successful epitaxy is matching lattice distances. If the spacing between atoms in the substrate is close to that of the top crystal, then that crystal will grow well; a small difference in lattice distance can be accommodated as the top crystal grows. When the lattice distances are different, however, the top crystal becomes deformed, since structural defects such as dislocations appear. Although few crystals share the same lattice distance, a number of examples are known. Aluminum arsenide and gallium arsenide have the same crystal structure and the same lattice parameters to within 0.1 percent; they grow excellent crystals on one another. Such materials, known as superlattices, have a repeated structure of n layers of GaAs, m layers of AlAs, n layers of GaAs, m layers of AlAs, and so forth. Superlattices represent artificially created structures that are thermodynamically stable; they have many applications in the modern electronics industry. Another lattice-matched epitaxial system is mercury telluride (HgTe) and cadmium telluride (CdTe). These two semiconductors form a continuous semiconductor alloy CdxHg1 " xTe, where x is any number between 0 and 1. This alloy is used as a detector of infrared radiation and is incorporated in particular in night-vision goggles.

Dendritic Growth

At slow rates of crystal growth, the interface between melt and solid remains planar, and growth occurs uniformly across the surface. At faster rates of crystal growth, instabilities are more likely to occur; this leads to dendritic growth. Solidification releases excess energy in the form of heat at the interface between solid and melt. At slow growth rates, the heat leaves the surface by diffusion. Rapid growth creates more heat, which is dissipated by convection (liquid flow) when diffusion is too slow. Convection breaks the planar symmetry so that crystal growth develops along columns, or "fingers," rather than along planes. Each crystal has certain directions in which growth is fastest, and dendrites grow in these directions. As the columns grow larger, their surfaces become flatter and more unstable. This feather or tree structure is characteristic of dendritic growth. Snowflakes are an example of crystals that result from dendritic growth.

ELECTRIC PROPERTIES

Resistivity

The German physicist Georg Simon Ohm discovered the basic law of electric conduction, which is now called Ohm's law. His law relates the voltage (V, measured in volts), the current (I, in amperes), and the resistance (R, in ohms) according to the formula V = RI. A current I through a solid induces a voltage V; the resistance R is the constant of proportionality. The value of R is an important factor in the design of electrical circuits. It is determined by the shape of the resistor: a long narrow object has more resistance than a short wide one of the same material. For solids, the important parameter is the resistivity ñ, which is given in units of ohm-metres. It is the resistivity per volume unit and is independent of shape. The relationship between R and ñ is R = ñL/A, where A is the area of the resistor and L is the length. These dimensions are measured in the direction of the current: L is the length of the current path, and A is the cross-sectional area. The resistance of a copper bar depends on its shape, but at a given temperature every piece of pure copper has the same resistivity. Thus the resistivity is a fundamental parameter of a material and is investigated by scientists. Resistivities of solids span a wide range of values. Certain metals have zero resistivity at low temperatures; they are called superconductors. At the other extreme, very good insulators such as sulfur and polystyrene have resistivities larger than one quadrillion ohm-metres. At room temperature, the metal with the lowest value of resistivity is silver, with

ñ = 1.6 × 10–8 ohm-metre; the second best conductor is copper, with ñ = 1.7 × 10–8 ohm-metre. Copper, rather than silver, is used in household wires because of the high cost of silver.

Conduction Through ion Hopping

Electrical conductivity ó is the inverse of resistivity and is measured in units of ohm-metre–1. Electrical current is produced by the motion of charges. In crystals, electrical current is due to the motion of both ions and electrons. Ions move by hopping occasionally from site to site; all solids can conduct electricity in this manner. When the voltage is zero, there is no net current because the ions hop randomly in all directions. The imposition of a small voltage causes the ions to slightly favour one direction of motion, which leads to a net flow of charge in that direction; this constitutes an electrical current. The electricity conducted by this process is quite small and is usually negligible compared with that carried by the electrons. When an ion hops, it must migrate to a vacant site, which could be either an interstitial or a vacancy. Ionic conductivity can occur because hopping ions cause vacancies to move through the solid. An ion hops to the vacant site, thereby filling the vacancy, while creating a new one at the ion's former site. Repeating this process causes the position of the vacancy to migrate through the crystal. The motion of the vacancy arises from the motion of ions, which carry charge and contribute to electrical conductivity.

Ion hops are induced by thermal fluctuations. Most of the ions move within their lattice site, vibrating around this point. Temperature is defined as the average energy of this vibrational motion; the more the ions move, the higher the temperature. An individual ion at times moves slowly and at times vibrates quite rapidly but usually has an energy near the average value. Each ion shares its vibrational energy with its neighbouring ions. An ion typically has some neighbours with small vibrations and others with large ones. The average energy shared with the neighbours is close to the average energy of all the atoms. As a random process, however, it occasionally happens that all neighbours of an ion may have large vibrations, in which case the ion will acquire an unusually high energy. This energy may be high enough to cause it to leave its site and hop to a neighbouring site. A thermal fluctuation is the rare process in which the energy at a local site may be much higher or lower than the average energy in the crystal. Probability theory shows that the higher the temperature, the more frequent are these thermal fluctuations. Ions therefore hop more often at high temperature.

A few solids conduct electricity better by ion motion than by electron motion. These unusual materials are technologically important in making batteries. All batteries have two electrodes separated by an electrolyte, which is a material that conducts ions better than electrons. An example of a crystal electrolyte is â-alumina, which readily conducts monovalent cations such as silver (Ag^+) and sodium (Na^+).

Among all ions, silver has the largest value of ionic conductivity in many different electronic insulators.

The copper ion (Cu+) forms the same type of chemical bonds as does the silver ion, but the copper ion, because of its smaller radius, does not migrate as well within an electrolyte. Silver ions fit perfectly into the interstitial sites of the crystal lattices of several electrolytes, while the smaller copper ions permit the neighbouring ions to collapse around them, inhibiting further hopping. There are a few good conductors of the inexpensive copper ion that can be used as solid electrolytes in batteries. Silver is too costly and heavy to use in large-volume batteries such as those found in automobiles, but it is used in the smaller batteries that power devices such as hearing aids.

Conduction Electrons

Electrons carry the basic unit of charge e, equal to 1.6022 × 10–19 coulomb. They have a small mass and move rapidly. Most electrons in solids are bound to the atoms in local orbits, but a small fraction of the electrons are available to move easily through the entire crystal. These so-called conduction electrons carry the electrical current. Solids with many conduction electrons are metals, while those with a few are semimetals or semiconductors. In insulators, nearly all the electrons are bound, and very few electrons are capable of carrying current. A typical metal has one or more conduction electrons in each atomic unit cell, a semiconductor may have only one conduction electron for each thousand unit cells, and an insulator may have one conduction electron per one million or one trillion unit cells.

The bonding properties of the individual atoms of a solid determine the behaviour of the bulk solid. The electrical properties of a solid can usually be predicted from the valence and bonding preferences of its atoms. In the argon atom, for example, all atomic shells are filled with electrons. The electrons of solid argon remain in the atomic shells; none are conduction electrons, and the electrical resistivity is therefore high. Solid argon, like all the rare gas solids, is a good insulator. A few conduction electrons are contributed by impurities, and so the conductivity, though small, is not zero. These conduction electrons move quite readily through the solid. The term mobility is used to describe how well a conduction electron moves through the solid in response to a voltage. Conductivity is the product of mobility, the electrical charge e, and the number N of conduction electrons per unit volume: ó = Neì, where ó is the conductivity and ì is the mobility. The mobility of the rare gas solids is high, but their conductivity is nonetheless low because there is a small number of conduction electrons.

Electrical Insulators

Like the rare gas solids, most ionic solids are electrical insulators. In sodium chloride, for example, each sodium atom donates its single valence electron to a chlorine atom, thus forming a solid composed of Na+ and Cl" ions. All electrons are in filled shells at low temperature, and in a perfect crystal there are no conduction electrons. Sodium chloride is thus an insulator with a very high resistivity. Some conduction electrons are provided by impurities or thermal excitations. At high temperatures large ion vibrations from thermal fluctuations may knock an electron out of a filled shell, upon which it becomes a conduction electron and contributes to the conductivity. The number of conduction electrons created by thermal excitations is small for most insulators. Although defects can be responsible for producing conduction electrons, they can also destroy the conducting ability of electrons by trapping them. The defects have local orbitals that provide a lower energy state for the electron than the one occupied in the conduction state. A conduction electron becomes bound at the defect, ceasing to contribute to the conductivity. This process is very efficient in insulators, so the few conduction electrons provided by impurities and thermal fluctuations are usually trapped at other defects. By definition, an insulator is a solid that does not provide a stable environment for conduction electrons.

CONDUCTIVITY OF METALS

Metals have a high density of conduction electrons. The aluminum atom has three valence electrons in a partially filled outer shell. In metallic aluminum the three valence electrons per atom become conduction electrons. The number of conduction electrons is constant, depending on neither temperature nor impurities. Metals conduct electricity at all temperatures, but for most metals the conductivity is best at low temperatures. Divalent atoms, such as magnesium or calcium, donate both valence electrons to become conduction electrons, while monovalent atoms, such as lithium or gold, donate one. As will be recalled, the number of conduction electrons alone does not determine conductivity; it depends on electron mobility as well. Silver, with only one conduction electron per atom, is a better conductor than aluminum with three, for the higher mobility of silver compensates for its fewer electrons.

In metals such as sodium and aluminum, the atoms donate all their valence electrons to the conduction band. The resulting ions are small, occupying only 10–15 percent of the volume of the crystal. The conduction electrons are free to roam through the remaining space. A simple model, which often describes well the properties of the conduction electrons, treats them as interacting neither with the ions nor with each other. The electrons are approximated as free particles wandering easily through the crystal. This concept was first proposed by the German scientist Arnold Johannes Wilhelm Sommerfeld. It works quite well for those metals, known as simple metals, whose conduction electrons are donated from sp-shells—for example, aluminum, magnesium, calcium, zinc, and lead. They are called simple because they are aptly described by the simple theory of Sommerfeld.

The transition metals are found in three rows of the periodic table: the first row consists of scandium through nickel, the second row is yttrium through palladium, and the third row is lanthanum plus hafnium through platinum. Within these rows, as the atomic number increases, the electrons fill d-states in the outer shell of the atom. In crystal form the transition metal atoms are metals with interesting properties. The d-electrons are more tightly bound to the ion centre than are sp-electrons. While the sp-valence electrons become conduction electrons that move freely through the crystal, the d-electrons tend to stay localized near the ion. Neighbouring ions may covalently bond d-electrons. In most cases, these d-states are only partially filled. Electrons in these d-states can conduct as well as those in the sp-states, but the electron motion in the d-states is not well approximated by the Sommerfeld model of free particles. Instead, the electrons move from ion to ion through the shared covalent bonds of the d-electrons. These metals have some conduction electrons donated from sp-states and others from d-states; therefore, some electrons move freely according to the Sommerfeld model, while others move through the bonds. Each electron switches back and forth between these two modes of conduction, resulting in electron motion that is quite complicated.

An applied voltage causes the electrons of metals to accelerate and contribute to the electric current. The electrons scatter occasionally from imperfections in the crystal, and the rate of scattering determines the mobility. The electrons do not scatter from the ions in the crystal that are located at the expected site in the crystal lattice. The electrons move to accommodate the host ions rather than scatter from them. If an ion is missing, misplaced, or of a different species, however, the electron will scatter from this defect. Ions vibrate around their lattice site, with the amplitude vibration increasing with temperature. The vibration may cause the ion to be displaced from its crystal site, providing a defect from which an electron will scatter. The resistivity of metals increases at high temperature, owing to the increase in vibrations of the ions in the crystal and the resulting increase in scattering.

Conducting Properties of Semiconductors

Semiconductors have conducting properties intermediate to those of insulators and metals. In some cases the semiconductors are insulators, while in others they are metals. Semiconductors share with insulators the property that they have no conduction electrons in a perfect crystal without thermal fluctuations. Conduction electrons are provided by electrons from impurities or by thermal fluctuation of electrons from atomic shells. The important difference between insulators and semiconductors is in the nature of the traps. A trap is a local electron energy state at a defect. Although the traps in insulators bind conduction electrons tightly, those in semiconductors only weakly bind the electrons. A trapped conduction electron in a semiconductor can be kicked back to the conduction band by thermal fluctuations. At room temperature, the majority of extra electrons are found in the conduction band rather than in traps. The inability of traps to keep electrons is the main difference between semiconductors and insulators. A semiconductor at room temperature has a sufficient number of conduction electrons to provide good electrical conductivity. Since the mobility of electrons in many semiconductors is exceptionally high, even a small number of conduction electrons is generally sufficient to allow high conductivity.

Phosphorus has five valence electrons, while silicon has four. When a phosphorus atom substitutes for an atom in a silicon crystal lattice, four of its five valence electrons enter covalent bonds. The fifth one is extra, sitting in a shallow trap around the phosphorus site. It is easily excited, however, to the conduction band by thermal fluctuations. At room temperature, there is nearly one conduction electron in silicon for each phosphorus impurity. By controlling the number of impurities, it is possible to control the conductivity of silicon. Other substitutional atoms such as arsenic and antimony also serve as electron donors to the conduction band of silicon.

If a sufficient number of conduction electrons are added to a semiconductor through the introduction of impurities, the electrical properties become metallic. There is a critical concentration of impurities Nc, which depends on the type of impurity. For impurity concentrations less than the critical amount Nc, the conduction electrons become bound in traps at extremely low temperatures, and the semiconductor becomes an insulator. For a concentration of impurities higher than Nc, the conduction electrons are not bound in traps at low temperatures, and the semiconductor exhibits metallic conduction. For phosphorus impurities in silicon, Nc = 2 × 1018 impurities per cubic centimetre. Although this number seems large, it represents about one phosphorus atom for each 100,000 silicon atoms. On a percentage basis, a small number of phosphorus atoms will change silicon from an insulator to a metallic conductor. Other semiconductors have similar properties. In gallium arsenide the critical concentration of impurities for metallic conduction is 100 times smaller than in silicon.

Gallium atoms, like those of phosphorus, can be used as substitutional impurities in silicon. Each atom contributes three electrons to covalent bonds. Since four electrons are needed to complete a tetrahedral arrangement, there is one electron absent per gallium atom from a full set of covalent bonds. The missing electron is called a hole. Holes can move around the crystal in a process similar to the motion of ion vacancies, except in this case there is an electron vacancy. An electron from a nearby covalent bond can jump over and fill the empty electron state, thereby moving the hole to the neighbouring bond. The hole contributes a positive charge, since it is the absence of an electron. The mobility of holes in response to an external voltage is almost as high as the mobility of conduction electrons. A semiconductor may have a high density of impurities that cause holes, and a high electrical conductivity is created by their motion. A p-type semiconductor is one with a preponderance of holes; an n-type semiconductor has a preponderance of conduction electrons. The symbols p and n come from the sign of the charge of the particles: positive for holes and negative for electrons.

Thermal fluctuations can excite an electron out of a covalent bond, making it a conduction electron. The bond is left with a missing electron, which constitutes a hole. Thermal fluctuations thus make electron-hole pairs. Usually the electron and hole separate in space, and each wanders away. The Swiss-American scientist Gregory Hugh Wannier first suggested that the electron and hole could bind together weakly. This bound state, called a Wannier exciton, does exist; the hole has a positive charge, the electron has a negative charge, and the opposites attract. The exciton is observed easily in experiments with electromagnetic radiation. It lives for only a short time—between a nanosecond and a microsecond—depending on the semiconductor. The short lifetime is due to the preference for the electron to reenter a covalent bond state, thereby eliminating both the hole and the conduction electron. This recombination of electron and hole is easily accomplished from the exciton state, since the two particles are spatially nearby. If the electron and hole escape the exciton state by thermal fluctuation, they travel away from each other. Recombination is then less probable, since it occurs only when the wandering particles pass close to one another again. Recombination also can occur at defect sites. First, one particle becomes bound to the defect, followed by the second particle. The electron and hole are again close to one another, and the electron can reoccupy the covalent bond.

As in metals, the mobility of electrons in semiconductors is limited by electron scattering. For crystals with few defects, the mobility is limited by defect scattering at the lowest temperatures and by ion vibrations at moderate and high temperatures.

Since semiconductors with few defects have a small number of conduction electrons, the resistivity is high. The number of conduction electrons is increased in semiconductors by adding impurities. Unfortunately, this also increases the scattering from impurities, which reduces the mobility. The resistivity of silicon at room temperature (T = 300 K) as a function of the concentration of impurities. The two curves represent conduction by electrons and by holes. Each grid mark on the graph is a factor of 10. The resistivity varies by a factor of one million from the lowest to the highest concentration of impurities.

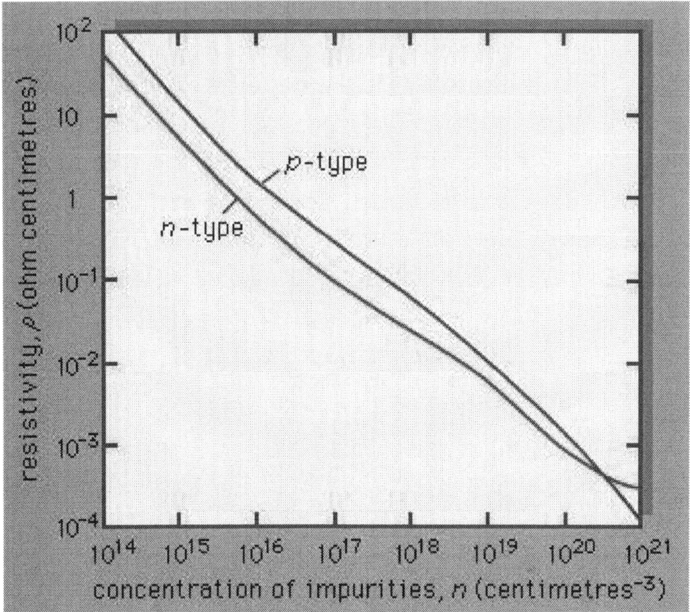

Fig.: The electrical resistivity (p) of a Silicon Semiconductor at a temperature of 300k as a function of the concentration of impurities (n)

Semiconductors with few impurities are good photoconductors. Photoconductivity is the phenomenon in which the electrical conductivity of a solid is increased by exposing it to light. Light is electromagnetic radiation within a specific narrow band of frequencies. The quanta of light are absorbed by the semiconductor, creating electron-hole pairs that provide the electrical conduction. More intense light produces more electron-hole pairs and gives rise to better conductivity. Each semiconductor absorbs light over a specific frequency range, so different semiconductors are used as photoconductors for different ranges of frequency.

Zinc oxide (ZnO) is an interesting material with respect to conductivity. It crystallizes in the wurtzite structure, and its bonding is a mix of ionic and covalent.

High-purity single crystals are insulators. Zinc oxide is the most piezoelectric of all materials and is widely used as a transducer in electronic devices. (Piezoelectricity is the property of a crystal to become polarized when subjected to pressure.) Zinc oxide is a good semiconductor when aluminum impurities are included in the crystal. Polycrystalline ceramics of semiconducting zinc oxide conduct well and obey Ohm's law. The addition of small amounts of other oxides, such as those of barium and chromium, causes zinc oxide ceramics to have very nonohmic electrical properties; the electrical current in such ceramics is the most nonlinear of any known material. The current I becomes proportional to a power of the voltage Vn, where the exponent n has values of more than 100 in certain ranges of voltage. This material is called a varistor, which is a contraction of the words variable and resistor. Zinc oxide varistors are widely used as circuit elements to protect against voltage surges. There is little current until a critical voltage of about 330 volts is reached, at which point the current rises steeply in a nonlinear fashion. Another interesting application of zinc oxide was its former use as a white pigment in paint. It has been replaced by titanium dioxide (TiO2), however, which is whiter.

MAGNETISM

Explanation of Magnetism

Electrons are perpetually rotating, and, since the electron has a charge, its spin produces a small magnetic moment. Magnetic moments are small magnets with north and south poles. The direction of the moment is from the south to the north pole. In nonmagnetic materials the electron moments cancel, since there is random ordering to the direction of the electron spins. Whenever two electrons have their moments aligned in opposite directions, their effects tend to cancel. Magnets are formed when a large number of the electrons align their individual moments in the same direction. Only a small percentage of crystals are magnetic. The forces that tend to align the electron spins are subtle. There are three separate parts to the explanation of magnetism.

1. Most magnets are composed of atoms whose valence electrons are in d- or f-shells. Atomic shell notation refers to angular momentum, where s has zero unit, p has one, d has two, and f has three. Electrons in d-shells tend to be bound to the ion, and those in f-shells are bound even more tightly.
2. Each electron orbital can be occupied by two electrons—one with spin up and one with spin down. The d-shell has five orbital states and 10 electrons when filled; the f-shell has seven orbital states and 14 electrons when filled. Electrons are added one at a time to the d-states according to the empirical

rule that the electrons arrange themselves in the state with the maximum spin and the maximum magnetic moment. If the first electron has spin up, the next four will also have spin up. A maximum of five electrons with spin up are allowed in the d-shell, so the sixth must have spin down. Similarly, the f-shell accepts seven electrons with the same direction of spin before taking electrons with the opposite spin orientation. The order in which electrons fill atomic shells is described by Hund's rules, of which the first is maximizing the total spin. Atoms with electrons in partially filled d- and f-shells usually have a nonzero total spin and thus a net magnetic moment. These magnetic ions are the building blocks for magnetic crystals. Hund's first rule is due to a phenomenon called electron exchange. As discussed above, a fundamental rule of quantum mechanics, the Pauli exclusion principle, states that no two electrons with the same direction of spin can occupy the same point in space at the same time. Electrons have charge and repel one another. If two electrons come close together, a large amount of repulsive energy is produced. Physical systems prefer the state of lowest energy, and so electrons avoid such close approach. When their spins are parallel, electrons avoid each other because of the Pauli principle. Electrons in the same shell thus prefer to have their spins parallel, since this configuration keeps the electrons apart and thereby reduces the amount of repulsive energy. The concept of electron exchange is the basis of magnetism. It explains why ions such as iron have large magnetic moments. In divalent iron (Fe2+) the six d-electrons are arranged to achieve maximum electron spin and magnetic moment.

3. Individual ions with fixed magnetic moments may cooperatively align their moments, resulting in the presence of magnetic properties of the crystal as a whole. Ferromagnet crystals have the magnetic moments from all their constituent ions aligned in the same direction; the magnetic moment of the crystal is the summation of the individual moments of the ions. There must be a magnetic force between the different ions that causes them to cooperatively align their moments. This force is also due to electron exchange. The d-orbitals from neighbouring ions overlap weakly into covalent bonds. The d-electrons on the separate ions are shared with the neighbour through covalent bonding. The electron exchange will tend to align the spins on the two neighbours. Aligning all pairs of neighbours aligns all ions. The exchange force between neighbours is much weaker than the force within the atomic shell of one ion. Although weak, the force is sufficient to cause ferromagnetism.

Ferromagnetic Materials

Iron is a typical ferromagnet. Not all bars of iron are magnets; the existence of magnetism is determined by the nature of the domains within the bar. A domain is a region of a crystal in which all the ions are ferromagnetically aligned in the same direction. A bar may be composed of many domains, each having a different magnetic orientation. Such a bar would not appear to be magnetic. Each piece of the bar is magnetic, but the domains have moments that point in different directions, so the bar has no net moment. If the bar of iron is placed in a strong magnetic field, however, the bar becomes magnetic. The field causes the bar to become a single domain with all moments aligned along the external field. The domains do not rotate their moments; instead, the walls between domains move. The domain with a moment along the field grows, while the others become smaller. If removed from the magnetic field, the iron bar will remain magnetized for a considerable time period. Nearly all bars of iron are polycrystalline: they have many small grains of single crystals, which are packed together with random orientation. A grain could be a single domain, a domain could include many grains, or a large grain could have several domains.

Ferromagnetic materials change their magnetic ordering at a characteristic temperature Tc called the Curie temperature. The Curie temperatures for three common ferromagnetics—iron, cobalt, and nickel—are 1,043 K; 1,394 K; and 631 K, respectively. For temperatures below Tc the magnetic moments of the ions are aligned and the crystal is magnetic. For temperatures above Tc the crystal is not ferromagnetic, since the individual atomic moments are no longer aligned. Above Tc the moments have short-range order but not long-range order. Short-range order means there is local ordering. If a moment points in one direction, its neighbours have a tendency to point in the same direction. This tendency is maintained over several lattice sites but is not maintained for long distances. Long-range order is the tendency for moments to align for large distances. For temperatures a few degrees below Tc the moments have strong short-range order but only a small amount of long-range order, so the bar is not very magnetic. The tendency for long-range order increases at lower temperature. The Curie temperature is the point where long-range order begins as the temperature is lowered.

If an iron bar is heated to a temperature above Tc, the bar is no longer magnetic. If the bar is then cooled to a temperature below Tc, the grains become magnetic, but they orient their moments in random directions, so the bar as a whole is not magnetic. A bar can be demagnetized by heating the bar and then cooling it. By inserting it in a large magnetic field, the bar can be remagnetized.

Ferromagnetism is found in many insulators as well as metals. Chromium bromide (CrBr3) is an insulator since chromium is trivalent and a bromine atom needs one electron to complete its outer shell. The trivalent chromium atoms each have a moment, and these align ferromagnetically below the Curie temperature of 37 K. Gadolinium chloride (GdCl3; Tc = 2.2 K) and europium oxide (EuO; Tc = 77 K) are two other examples among many.

Antiferromagnetic Materials

Many crystals have magnetic ions that are ordered in arrangements other than ferromagnetic. In antiferromagnetic ordering, the moments pointing in one direction are balanced by others pointing in the opposite direction, with the result that the substance has no net magnetization. The exchange interaction between ions in this case has the opposite sign and favours the alternate arrangements of spins. The sign of the exchange interaction between ions depends on the length of the covalent bond and the bonding angles; it may have either orientation. The characteristic temperature associated with antiferromagnetism is called the Néel temperature TN. Below TN the ions are antiferromagnetically ordered, while above this temperature there is no long-range antiparallel order. Some examples of antiferromagnetic crystals are manganese oxide (MnO; TN = 116 K), manganese sulfide (MnS; TN = 160 K), and iron oxide (FeO; TN = 198 K). Manganese oxide is an insulator since manganese atoms are divalent and oxygen atoms accept two electrons. The manganese ion has a fixed magnetic moment. The crystal structure of manganese oxide is the same as that of sodium chloride shown in Figure.

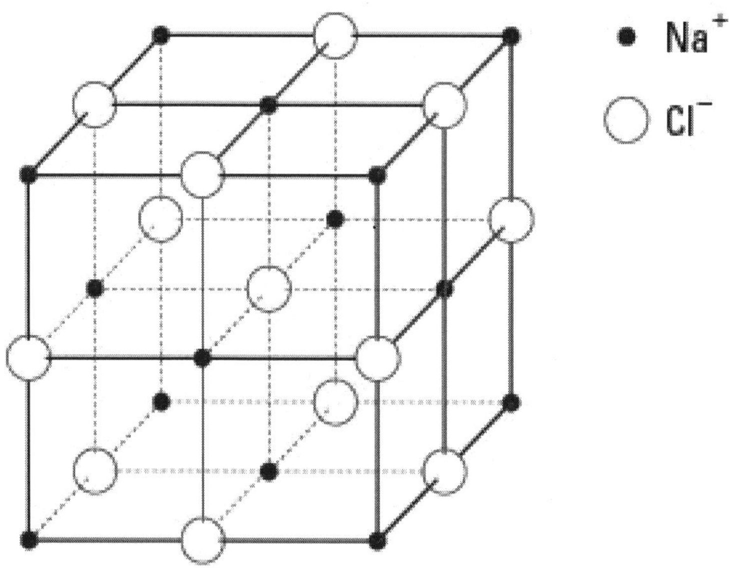

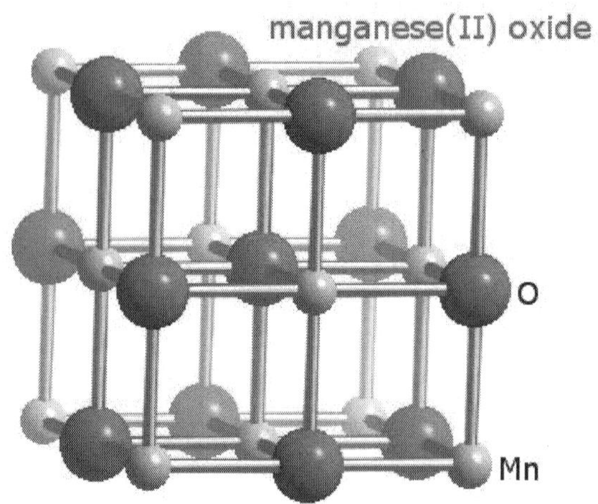

Below the Néel temperature the atomic unit cell doubles in size to include two atoms of each type of ion. This is necessary because below TN neighbouring manganese atoms have moments in the opposite direction and are no longer equivalent; the unit cell must therefore include one moment in each of the two directions. Fluorides such as manganese fluoride (MnF2), iron (I) fluoride (FeF2), cobalt fluoride (CoF2), and nickel fluoride (NiF2) are other crystals that exhibit antiferromagnetic ordering of the transition metal ions.

Ferrimagnetic Materials

Ferrimagnetism is another type of magnetic ordering. In ferrimagnets the moments are in an antiparallel alignment, but they do not cancel. The best example of a ferrimagnetic mineral is magnetite (Fe3O4). Two iron ions are trivalent, while one is divalent. The two trivalent ions align with opposite moments and cancel one another, so the net moment arises from the divalent iron ion. The historic lodestone that was the first magnetic material discovered was a form of magnetite. Another class of ferrimagnets has the garnet structure; Y3Fe5O12 is one such crystal. Only the iron ions are magnetic. Three point in one direction and two in the other, so there is a net magnetic moment of one iron ion in the unit cell. However, if a rare-earth ion such as gadolinium (Gd) is substituted for yttrium (Y), then the rare-earth ion also contributes to the ferrimagnetism.

Ferrites are oxides of iron with the formula MFe2O4, where M is a divalent ion such as nickel, zinc, cadmium, manganese, or magnesium.

The ferric iron ion is trivalent, and the oxygen ion accepts two electrons. Actually M can also be divalent iron, forming magnetite (Fe3O4). The crystal structure is called spinel, which is the mineral name for MgAl2O4. Ferrites are electrical insulators with magnetic ordering. Their insulating quality makes them useful as magnetic cores. When metallic ferromagnetic materials are exposed to alternating magnetic fields, significant heating losses occur from eddy currents. Ferrite magnets greatly reduce such heat losses because of their high resistivity. They also absorb electromagnetic radiation of very long wavelengths. This property is unusual, since metals reflect such radiation while insulators transmit it. In small crystallites of ferrites, the electromagnetic radiation causes the magnetic moment to rotate at the frequency of light. The absorption frequency of the light depends on the detailed shape of the crystal grain. A polycrystalline material, with a range of grain sizes and shapes, absorbs over a broad range of radar frequencies. Ferrite crystals are a major ingredient in snooper paint, which makes stealth airplanes undetectable by radar. Regular airplanes reflect radar, and the reflected signals can be used to locate the aircraft. Stealth planes absorb the radar signals and so cannot be located in this way.

The Kondo Effect

Magnetic ions have interesting properties when they are found as impurities in nonmagnetic crystals. They usually retain their magnetic moment, so small magnets are distributed randomly throughout the crystal. If the host crystal is a metal, the magnetic impurities make an interesting contribution to the electrical resistivity. The conduction electrons scatter from the magnetic impurity. Since the conduction electron and the impurity both have spin, they can mutually flip spins while scattering. The spin-flip scattering is strong at low temperatures and actually increases slightly as temperature decreases. This phenomenon is called the Kondo effect after the Japanese theoretical physicist Jun Kondo, who first explained the increase in resistivity resulting from magnetic impurities. There is a characteristic temperature, called the Kondo temperature, which depends on the impurity and on the metallic host. The resistivity increases at low temperature, starting near the Kondo temperature. A typical example of a Kondo system is iron impurities in copper; the system's Kondo temperature is 24 K.

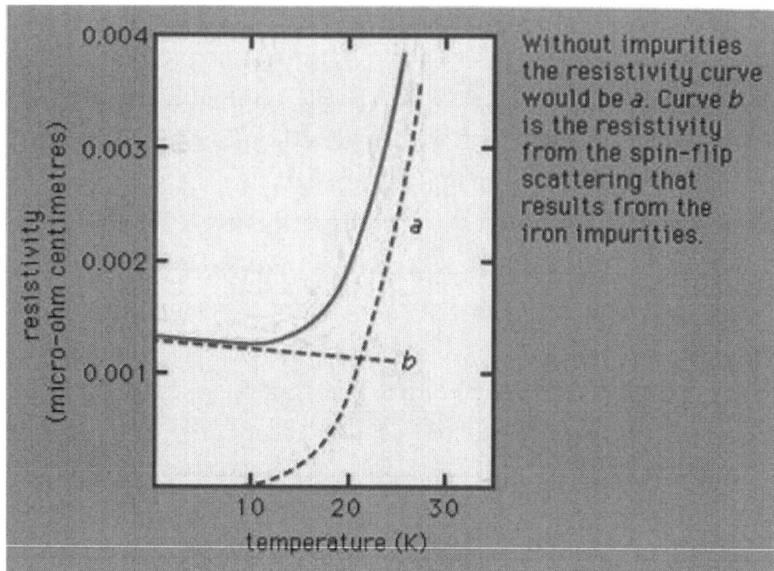

The solid line in Figure shows the resistivity in copper at low temperature when there are 110 iron impurities per 1,000,000 copper atoms. The dashed line a is the resistivity in the absence of impurities. It increases at higher temperature because the electron scatters from ion vibrations. The dashed line b is the resistivity from spin-flip scattering. Nearly all transition metal atoms are found as magnetic impurities in copper or gold. Each system has a different Kondo temperature, which varies from 1,000 K to a fraction of 1 K. The spin-flip part of the electrical resistivity is unique in that it is large at low temperatures and decreases at high temperatures; most contributions to the resistivity increase at high temperatures.

8

SOUND

In physics, sound is a vibration that propagates as a typically audible mechanical wave of pressure and displacement, through a medium such asair or water. In physiology and psychology, sound is the reception of such waves and their perception by the brain. Acoustics

Acoustics is the interdisciplinary science that deals with the study of mechanical waves in gases, liquids, and solids including vibration, sound, ultrasound, and infrasound. A scientist who works in the field of acoustics is an acoustician, while someone working in the field of acoustical engineering may be called an acoustical engineer. An audio engineer, on the other hand is concerned with the recording, manipulation, mixing, and reproduction of sound.

Applications of acoustics are found in almost all aspects of modern society, subdisciplines include aeroacoustics, audio signal processing, architectural acoustics, bioacoustics, electro-acoustics, environmental noise, musical acoustics, noise control, psychoacoustics, speech, ultrasound, underwater acoustics, and vibration.

DEFINITION

Sound is defined by ANSI/ASA S1. 1-2013 as "(a) Oscillation in pressure, stress, particle displacement, particle velocity, etc., propagated in a medium with internal forces (e. g., elastic or viscous), or the superposition of such propagated oscillation. (b) Auditory sensation evoked by the oscillation described in (a). "

Physics of sound

Sound can propagate through compressible media such as air, water and solids as longitudinal waves and also as a transverse waves in solids. The sound waves are generated by a sound source, such as the vibrating diaphragm of a stereo speaker. The sound source creates vibrations in the surrounding medium. As the source continues to vibrate the medium, the vibrations propagate away from the source at the speed of sound, thus forming the sound wave. At a fixed distance from the source, the pressure, velocity, and displacement of the medium vary in time. At an instant in time, the pressure, velocity, and displacement vary in space. Note that the particles of the medium do not travel with the sound wave. This is intuitively obvious for a solid, and the same is true for liquids and gases (that is, the vibrations of particles in the gas or liquid transport the vibrations, while the average position of the particles over time does not change). During propagation, waves can be reflected, refracted, or attenuated by the medium.

The behavior of sound propagation is generally affected by three things:

- A relationship between density and pressure. This relationship, affected by temperature, determines the speed of sound within the medium.
- The propagation is also affected by the motion of the medium itself. For example, sound moving through wind. Independent of the motion of sound through the medium, if the medium is moving, the sound is further transported.
- The viscosity of the medium also affects the motion of sound waves. It determines the rate at which sound is attenuated. For many media, such as air or water, attenuation due to viscosity is negligible.

When sound is moving through a medium that does not have constant physical properties, it may be refracted (either dispersed or focused). The mechanical vibrations that can be interpreted as sound are able to travel through all forms of matter: gases, liquids, solids, andplasmas. The matter that supports the sound is called the medium. Sound cannot travel through a vacuum.

Longitudinal and transverse waves

Sound is transmitted through gases, plasma, and liquids as longitudinal waves, also called compression waves. It requires a medium to propagate. Through solids, however, it can be transmitted as both longitudinal waves and transverse waves. Longitudinal sound waves are waves of alternating pressure deviations from the equilibrium pressure, causing local regions of compression andrarefaction, while transverse waves (in solids) are waves of alternating shear stress at right angle to the direction of propagation.

Sound waves may be "viewed" using parabolic mirrors and objects that produce sound.

The energy carried by an oscillating sound wave converts back and forth between the potential energy of the extra compression (in case of longitudinal waves) or lateral displacement strain (in case of transverse waves) of the matter, and the kinetic energy of the displacement velocity of particles of the medium.

Sound wave properties and characteristics

Sound waves are often simplified to a description in terms of sinusoidal plane waves, which are characterized by these generic properties:

- Frequency, or its inverse, the period
- Wavelength
- Wave number
- Amplitude
- Sound pressure
- Sound intensity
- Speed of sound
- Direction

Sound that is perceptible by humans has frequencies from about 20 Hz to 20, 000 Hz. In air at standard temperature and pressure, the corresponding wavelengths of sound waves range from 17 m to 17 mm. Sometimes speed and direction are combined as a velocity vector; wave number and direction are combined as a wave vector.

Transverse waves, also known as shear waves, have the additional property, polarization, and are not a characteristic of sound waves.

Speed of sound

The speed of sound depends on the medium that the waves pass through, and is a fundamental property of the material. The first significant effort towards the measure of the speed of sound was made by Newton. He believed that the speed of sound in a particular substance was equal to the square root of the pressure acting on it (STP) divided by its density.

$$c = \sqrt{\frac{p}{\rho}}$$

This was later proven wrong when found to incorrectly derive the speed. French mathematician Laplace corrected the formula by deducing that the phenomenon of sound travelling is not isothermal, as believed by Newton, but adiabatic. He added another factor to the equation-gamma-and multiplied $\sqrt{\gamma}$ to $\sqrt{\frac{p}{\rho}}$, thus coming up with the equation $c=\sqrt{\gamma\cdot\frac{p}{\rho}}$. Since K = γ.p the final equation came up to be $c=\sqrt{\frac{K}{\rho}}$ which is also known as the Newton-Laplace equation. In this equation, K = elastic bulk modulus, c = velocity of sound, and = density. Thus, the speed of sound is proportional to the square root of the ratio of the bulk modulus of the medium to its density.

Those physical properties and the speed of sound change with ambient conditions. For example, the speed of sound in gases depends on temperature. In 20 °C (68 °F) air at sea level, the speed of sound is approximately 343 m/s (1, 230 km/h; 767 mph) using the formula "v = (331 + 0. 6 T) m/s". In fresh water, also at 20 °C, the speed of sound is approximately 1, 482 m/s (5, 335 km/h; 3, 315 mph). In steel, the speed of sound is about 5, 960 m/s (21, 460 km/h; 13, 330 mph). The speed of sound is also slightly sensitive (a second-order anharmonic effect) to the sound amplitude, which means that there are non-linear propagation effects, such as the production of harmonics and mixed tones not present in the original sound.

Perception of sound

A distinct use of the term sound from its use in physics is that in physiology and psychology, where the term refers to the subject of perception by the brain. The field of psychoacoustics is dedicated to such studies.

The physical reception of sound in any hearing organism is limited to a range of frequencies. Humans normally hear sound frequencies between approximately 20 Hz and 20, 000 Hz (20 kHz), The upper limit decreases with age.

Other species have a different range of hearing. For example, dogs can perceive vibrations higher than 20 kHz, but are deaf below 40 Hz. As a signal perceived by one of the major senses, sound is used by many species for detecting danger, navigation, predation, and communication. Earth'satmosphere, water, and virtually any physical phenomenon, such as fire, rain, wind, surf, or earthquake, produces (and is characterized by) its unique sounds. Many species, such as frogs, birds, marine and terrestrial mammals, have also developed special organs to produce sound. In some species, these produce song and speech. Furthermore, humans have developed culture and technology (such as music, telephone and radio) that allows them to generate, record, transmit, and broadcast sound.

Sometimes sound refers to only those vibrations with frequencies that are within the hearing range for humans or for a particular animal.

NOISE

Noise is a term often used to refer to an unwanted sound. In science and engineering, noise is an undesirable component that obscures a wanted signal.

Sound pressure level

Sound pressure is the difference, in a given medium, between average local pressure and the pressure in the sound wave. A square of this difference (i. e., a square of the deviation from the equilibrium pressure) is usually averaged over time and/or space, and a square root of this average provides a root mean square (RMS) value. For example, 1 Pa RMS sound pressure (94 dBSPL) in atmospheric air implies that the actual pressure in the sound wave oscillates between (1 atm Pa) and (1 atm Pa), that is between 101323. 6 and 101326. 4 Pa. As the human ear can detect sounds with a wide range of amplitudes, sound pressure is often measured as a level on a logarithmic decibelscale. The sound pressure level (SPL) or Lp is defined as

$$L_\mathrm{p} = 10 \log_{10}\left(\frac{p^2}{p_\mathrm{ref}{}^2}\right) = 20 \log_{10}\left(\frac{p}{p_\mathrm{ref}}\right) \mathrm{dB}$$

where p is the root-mean-square sound pressure and is a reference sound pressure. Commonly used reference sound pressures, defined in the standard ANSI S1. 1-1994, are 20 µPa in air and 1 µPa in water. Without a specified reference sound pressure, a value expressed in decibels cannot represent a sound pressure level.

Since the human ear does not have a flat spectral response, sound pressures are often frequency weighted so that the measured level matches perceived levels more closely. TheInternational Electrotechnical Commission (IEC) has defined several weighting schemes. A-weighting attempts to match the response of the human ear to noise and A-weighted sound pressure levels are labeled dBA. C-weighting is used to measure peak levels.

SPEED OF SOUND

History

Sir Isaac Newton computed the speed of sound in air as 979 feet per second (298 m/s), which is too low by about 15%, but had neglected the effect of fluctuating temperature; that was later rectified by Laplace.

During the 17th century, there were several attempts to measure the speed of sound accurately, including attempts by Marin Mersenne in 1630 (1, 380 Parisian feet per second), Pierre Gassendi in 1635 (1, 473 Parisian feet per second) and Robert Boyle (1, 125 Parisian feet per second).

In 1709, the Reverend William Derham, Rector of Upminster, published a more accurate measure of the speed of sound, at 1, 072 Parisian feet per second. Derham used a telescope from the tower of the church of St Laurence, Upminster to observe the flash of a distant shotgun being fired, and then measured the time until he heard the gunshot with a half second pendulum. Measurements were made of gunshots from a number of local landmarks, including North Ockendon church. The distance was known by triangulation, and thus the speed that the sound had travelled could be calculated.

Basic concept

The transmission of sound can be illustrated by using a model consisting of an array of balls interconnected by springs. For real material the balls represent molecules and the springs represent the bonds between them. Sound passes through the model by compressing and expanding the springs, transmitting energy to neighbouring balls, which transmit energy to their springs, and so on. The speed of sound through the model depends on the stiffness of the springs (stiffer springs transmit energy more quickly). Effects like dispersion and reflection can also be understood using this model.

In a real material, the stiffness of the springs is called the elastic modulus, and the mass corresponds to the density. All other things being equal (ceteris paribus), sound will travel more slowly in spongy materials, and faster in stiffer ones. For instance, sound will travel 1. 59 times faster in nickel than in bronze, due to the greater stiffness of nickel at about the same density. Similarly, sound travels about 1. 41 times faster in light hydrogen (protium) gas than in heavy hydrogen (deuterium) gas, since deuterium has similar properties but twice the density. At the same time, "compression-type" sound will travel faster in solids than in liquids, and faster in liquids than in gases, because the solids are more difficult to compress than liquids, while liquids in turn are more difficult to compress than gases.

Some textbooks mistakenly state that the speed of sound increases with increasing density. This is usually illustrated by presenting data for three materials, such as air, water and steel, which also have vastly different compressibilities which more than make up for the density differences. An illustrative example of the two effects is that sound travels only 4. 3 times faster in water than air, despite enormous differences in compressibility of the two media. The reason is that the larger density of water, which works to slow sound in water relative to air, nearly makes up for the compressibility differences in the two media.

Compression and shear waves

In a gas or liquid, sound consists of compression waves. In solids, waves propagate as two different types. A longitudinal wave is associated with compression and decompression in the direction of travel, which is the same process as all sound waves in gases and liquids. A transverse wave, called a shear wave in solids, is due to elastic deformation of the medium perpendicular to the direction of wave travel; the direction of shear-deformation is called the "polarization" of this type of wave. In general, transverse waves occur as a pair oforthogonal polarizations. These different waves (compression waves and the different polarizations of shear waves) may have different speeds at the same frequency. Therefore, they arrive at an observer at different times, an extreme example being an earthquake, where sharp compression waves arrive first, and rocking transverse waves seconds later.

The speed of a compression wave in fluid is determined by the medium's compressibility and density. In solids, the compression waves are analogous to those in fluids, depending on compressibility, density, and the additional factor of shear modulus. The speed of shear waves, which can occur only in solids, is determined simply by the solid material's shear modulus and density.

Equations

In general, the speed of sound c is given by the Newton-Laplace equation:

$$c = \sqrt{\frac{K_s}{\rho}},$$

where

- K_s is a coefficient of stiffness, the isentropic bulk modulus (or the modulus of bulk elasticity for gases);
- ρ is the density.

Thus the speed of sound increases with the stiffness (the resistance of an elastic body to deformation by an applied force) of the material, and decreases with the density. For ideal gases the bulk modulus K is simply the gas pressure multiplied by the dimensionless adiabatic index, which is about 1. 4 for air under normal conditions of pressure and temperature.

For general equations of state, if classical mechanics is used, the speed of sound is given by

$$c = \sqrt{\left(\frac{\partial p}{\partial \rho}\right)_s},$$

where

- ρ is the pressure;
- ρ is the density and the derivative is taken adiabatically, that is, at constant entropy s.

If relativistic effects are important, the speed of sound is calculated from the relativistic Euler equations.

In a non-dispersive medium, sound speed is independent of sound frequency, so the speeds of energy transport and sound propagation are the same for all frequencies. Air, a mixture of oxygen and nitrogen, constitutes a non-dispersive medium. However, air does contain a small amount of CO2 which is a dispersive medium, and causes dispersion to air at ultrasonic frequencies (> 28 kHz).

In a dispersive medium sound speed is a function of sound frequency, through the dispersion relation. Each frequency component propagates at its own speed, called the phase velocity, while the energy of the disturbance propagates at the group velocity. The same phenomenon occurs with light waves; see optical dispersion for a description.

Dependence on the properties of the medium

The speed of sound is variable and depends on the properties of the substance through which the wave is travelling. In solids, the speed of transverse (or shear) waves depends on the shear deformation under shear stress (called the shear modulus), and the density of the medium. Longitudinal (or compression) waves in solids depend on the same two factors with the addition of a dependence on compressibility.

In fluids, only the medium's compressibility and density are the important factors, since fluids do not transmit shear stresses. In heterogeneous fluids, such as a liquid filled with gas bubbles, the density of the liquid and the compressibility of the gas affect the speed of sound in an additive manner, as demonstrated in the hot chocolate effect.

In gases, adiabatic compressibility is directly related to pressure through the heat capacity ratio (adiabatic index), and pressure and density are inversely related at a given temperature and composition, thus making only the latter independent properties (temperature, molecular composition, and heat capacity ratio) important. In low molecular weightgases such as helium, sound propagates faster compared to heavier gases such as xenon (for monatomic gases the speed of sound is about 75% of the mean speed that molecules move in the gas). For a given ideal gas the sound speed depends only on its temperature. At a constant temperature, the ideal gas pressure has no effect on the speed of sound, because pressure and density (also proportional to pressure) have equal but opposite effects on the speed of sound, and the two contributions cancel out exactly. In a similar way, compression waves in solids depend both on compressibility and density—just as in liquids—but in gases the density contributes to the compressibility in such a way that some part of each attribute factors out, leaving only a dependence on temperature, molecular weight, and heat capacity ratio. Thus, for a single given gas (where molecular weight does not change) and over a small temperature range (where heat capacity is relatively constant), the speed of sound becomes dependent on only the temperature of the gas.

In non-ideal gases, such as a van der Waals gas, the proportionality is not exact, and there is a slight dependence of sound velocity on the gas pressure.

Humidity has a small but measurable effect on sound speed (causing it to increase by about 0. 1%-0. 6%), because oxygen and nitrogen molecules of the air are replaced by lighter molecules of water. This is a simple mixing effect.

ALTITUDE VARIATION AND IMPLICATIONS FOR ATMOSPHERIC ACOUSTICS

In the Earth's atmosphere, the chief factor affecting the speed of sound is the temperature. For a given ideal gas with constant heat capacity and composition, sound speed is dependent solely upon temperature; see Details below. In such an ideal case, the effects of decreased density and decreased pressure of altitude cancel each other out, save for the residual effect of temperature.

Since temperature (and thus the speed of sound) decreases with increasing altitude up to 11 km, sound is refracted upward, away from listeners on the ground, creating an acoustic shadow at some distance from the source. The decrease of the sound speed with height is referred to as a negative sound speed gradient.

However, there are variations in this trend above 11 km. In particular, in the stratosphere above about 20 km, the speed of sound increases with height, due to an increase in temperature from heating within the ozone layer. This produces a positive sound speed gradient in this region. Still another region of positive gradient occurs at very high altitudes, in the aptly-named thermosphere above 90 km.

Practical formula for dry air

The approximate speed of sound in dry (0% humidity) air, in meters per second, at temperatures near 0 °C, can be calculated from

$$c_{\text{air}} = (331.3 + 0.606 \cdot \vartheta)\ \text{m/s},$$

where is the temperature in degrees Celsius (°C).

This equation is derived from the first two terms of the Taylor expansion of the following more accurate equation:

$$c_{\text{air}} = 331.3\ \text{m/s}\sqrt{1 + \frac{\vartheta}{273.15}}.$$

Dividing the first part, and multiplying the second part, on the right hand side, by $\sqrt{273.15}$ gives the exactly equivalent form

$$c_{\text{air}} = 20.05\ \text{m/s}\sqrt{\vartheta + 273.15}.$$

The value of 331. 3 m/s, which represents the speed at 0 °C (or 273. 15 K), is based on theoretical (and some measured) values of the heat capacity ratio γ, as well as on the fact that at 1 atm real air is very well described by the ideal gas approximation. Commonly found values for the speed of sound at 0 °C may vary from 331. 2 to 331. 6 due to the assumptions made when it is calculated. If ideal gas is assumed to be 7/5 = 1. 4 exactly, the 0 °C speed is calculated to be 331. 3 m/s, the coefficient used above.

This equation is correct to a much wider temperature range, but still depends on the approximation of heat capacity ratio being independent of temperature, and for this reason will fail, particularly at higher temperatures. It gives good predictions in relatively dry, cold, low pressure conditions, such as the Earth's stratosphere. The equation fails at extremely low pressures and short wavelengths, due to dependence on the assumption that the wavelength of the sound in the gas is much longer than the average mean free path between gas molecule collisions. A derivation of these equations will be given in the following section.

A graph comparing results of the two equations is at right, using the slightly different value of 331. 5 m/s for the speed of sound at 0 °C.

DETAILS

Speed in ideal gases and in air

For an ideal gas, K (the bulk modulus in equations above, equivalent to C, the coefficient of stiffness in solids) is given by: K = γ.p.

thus, from the Newton-Laplace equation above, the speed of sound in an ideal gas is given by:

$$c = \sqrt{\gamma \cdot \frac{p}{\rho}},$$

where

- γ is the adiabatic index also known as the isentropic expansion factor. It is the ratio of specific heats of a gas at a constant-pressure to a gas at a constant-volume (C_p/C_v), and arises because a classical sound wave induces an adiabatic compression, in which the heat of the compression does not have enough time to escape the pressure pulse, and thus contributes to the pressure induced by the compression;
- p is the pressure;
- ρ is the density.

Using the ideal gas law to replace with nRT/V, and replacing ñ with nM/V, the equation for an ideal gas becomes

$$c_{\mathrm{ideal}} = \sqrt{\gamma \cdot \frac{p}{\rho}} = \sqrt{\frac{\gamma \cdot R \cdot T}{M}} = \sqrt{\frac{\gamma \cdot k \cdot T}{m}},$$

where

- C_{ideal} is the speed of sound in an ideal gas;
- R (approximately 8. 3145 J·mol·K) is the molar gas constant;
- k is the Boltzmann constant;
- γ (gamma) is the adiabatic index (sometimes assumed 7/5 = 1. 400 for diatomic molecules from kinetic theory, assuming from quantum theory a temperature range at which thermal energy is fully partitioned into rotation (rotations are fully excited), but none into vibrational modes. Gamma is actually experimentally measured over a range from 1. 3991 to 1. 403 at 0 °C, for air. Gamma is assumed from kinetic theory to be exactly 5/3 = 1. 6667 for monatomic gases such as noble gases);
- T is the absolute temperature;
- M is the molar mass of the gas. The mean molar mass for dry air is about 0. 0289645 kg/mol;
- m is the mass of a single molecule.

This equation applies only when the sound wave is a small perturbation on the ambient condition, and the certain other noted conditions are fulfilled, as noted below. Calculated values for C_{air} have been found to vary slightly from experimentally determined values.

Newton famously considered the speed of sound before most of the development of thermodynamics and so incorrectly used isothermal calculations instead of adiabatic. His result was missing the factor of but was otherwise correct.

Numerical substitution of the above values gives the ideal gas approximation of sound velocity for gases, which is accurate at relatively low gas pressures and densities (for air, this includes standard Earth sea-level conditions). Also, for diatomic gases the use of requires that the gas exists in a temperature range high enough that rotational heat capacity is fully excited (i. e., molecular rotation is fully used as a heat energy "partition" or reservoir); but at the same time the temperature must be low enough that molecular vibrational modes contribute no heat capacity (i. e., insignificant heat goes into vibration, as all vibrational quantum modes above the minimum-energy-mode, have energies too high to be populated by a significant number of molecules at this temperature). For air, these conditions are fulfilled at room temperature, and also temperatures considerably below room temperature. See the section on gases in specific heat capacity for a more complete discussion of this phenomenon.

For air, we use a simplified symbol.

Additionally, if temperatures in degrees Celsius(°C) are to be used to calculate air speed in the region near 273 kelvin, then Celsius temperature may be used. Then

$$c_{\mathrm{ideal}} = \sqrt{\gamma \cdot R_* \cdot T} = \sqrt{\gamma \cdot R_* \cdot (\vartheta + 273.15)}, \quad c_{\mathrm{ideal}} = \sqrt{\gamma \cdot R_* \cdot 273.15} \cdot \sqrt{1 + \frac{\vartheta}{273.15}}.$$

For dry air, where ϑ(theta) is the temperature in degrees Celsius(°C).

Making the following numerical substitutions, R = 8.314510 J/(mol.K)

is the molar gas constant in J/mole/Kelvin, and M_{air} = 0.0289645 kg/mol

is the mean molar mass of air, in kg; and using the ideal diatomic gas value of γ = 1.4000.

$$\text{Then } c_{\mathrm{air}} = 331.3\ \mathrm{m/s}\sqrt{1 + \frac{\vartheta\,{}^\circ\mathrm{C}}{273.15\ {}^\circ\mathrm{C}}}.$$

Using the first two terms of the Taylor expansion:

$$c_{\mathrm{air}} = 331.3\ \mathrm{m/s}\left(1 + \frac{\vartheta\,{}^\circ\mathrm{C}}{2 \cdot 273.15\ {}^\circ\mathrm{C}}\right). \quad c_{\mathrm{air}} = (331.3 + 0.606\ {}^\circ\mathrm{C}^{-1} \cdot \vartheta)\ \mathrm{m/s}.$$

The derivation includes the first two equations given in the Practical formula for dry air section above.

Effects due to wind shear

The speed of sound varies with temperature. Since temperature and sound velocity normally decrease with increasing altitude, sound is refracted upward, away from listeners on the ground, creating an acoustic shadow at some distance from the source. Wind shear of 4 m·s·km can produce refraction equal to a typical temperature lapse rate of7. 5 °C/km. Higher values of wind gradient will refract sound downward toward the surface in the downwind direction, eliminating the acoustic shadow on the downwind side. This will increase the audibility of sounds downwind. This downwind refraction effect occurs because there is a wind gradient; the sound is not being carried along by the wind.

For sound propagation, the exponential variation of wind speed with height can be defined as follows:

$$U(h) = U(0)h^{\zeta}\frac{dU}{dH} = \zeta\frac{U(h)}{h},$$

where

where

- U(h) = speed of the wind at height, and is a constant;
- ζ = exponential coefficient based on ground surface roughness, typically between 0. 08 and 0. 52;
- $\frac{dU}{dH}$ = expected wind gradient at height.

In the 1862 American Civil War Battle of Iuka, an acoustic shadow, believed to have been enhanced by a northeast wind, kept two divisions of Union soldiers out of the battle, because they could not hear the sounds of battle only 10 km (six miles) downwind.

Tables

In the standard atmosphere:

- T0 is 273. 15 K (= 0 °C = 32 °F), giving a theoretical value of 331. 3 m/s (= 1086. 9 ft/s = 1193 km/h = 741. 1 mph = 644. 0 kn). Values ranging from 331. 3-331. 6 may be found in reference literature, however;
- T20 is 293. 15 K (= 20 °C = 68 °F), giving a value of 343. 2 m/s (= 1126. 0 ft/s = 1236 km/h = 767. 8 mph = 667. 2 kn);
- T25 is 298. 15 K (= 25 °C = 77 °F), giving a value of 346. 1 m/s (= 1135. 6 ft/s = 1246 km/h = 774. 3 mph = 672. 8 kn).

In fact, assuming an ideal gas, the speed of sound c depends on temperature only, not on the pressure or density (since these change in lockstep for a given temperature and cancel out). Air is almost an ideal gas. The temperature of the air varies with altitude, giving the following variations in the speed of sound using the standard atmosphere—actual conditions may vary.

EFFECT OF FREQUENCY AND GAS COMPOSITION

General physical considerations

The medium in which a sound wave is travelling does not always respond adiabatically, and as a result the speed of sound can vary with frequency.

The limitations of the concept of speed of sound due to extreme attenuation are also of concern. The attenuation which exists at sea level for high frequencies applies to successively lower frequencies as atmospheric pressure decreases, or as the mean free path increases. For this reason, the concept of speed of sound (except for frequencies approaching zero) progressively loses its range of applicability at high altitudes. The standard equations for the speed of sound apply with reasonable accuracy only to situations in which the wavelength of the soundwave is considerably longer than the mean free path of molecules in a gas.

The molecular composition of the gas contributes both as the mass (M) of the molecules, and their heat capacities, and so both have an influence on speed of sound. In general, at the same molecular mass, monatomic gases have slightly higher sound speeds (over 9% higher) because they have a higher (5/3 = 1. 66...) than diatomics do (7/5 = 1. 4). Thus, at the same molecular mass, the sound speed of a monatomic gas goes up by a factor of

$$\frac{c_{\text{gas,monatomic}}}{c_{\text{gas,diatomic}}} = \sqrt{\frac{5/3}{7/5}} = \sqrt{\frac{25}{21}} = 1.091\ldots$$

This gives the 9% difference, and would be a typical ratio for sound speeds at room temperature in helium vs. deuterium, each with a molecular weight of 4. Sound travels faster in helium than deuterium because adiabatic compression heats helium more, since the helium molecules can store heat energy from compression only in translation, but not rotation. Thus helium molecules (monatomic molecules) travel faster in a sound wave and transmit sound faster. (Sound generally travels at about 70% of the mean molecular speed in gases).

Note that in this example we have assumed that temperature is low enough that heat capacities are not influenced by molecular vibration. However, vibrational modes simply cause gammas which decrease toward 1, since vibration modes in a polyatomic gas gives the gas additional ways to store heat which do not affect temperature, and thus do not affect molecular velocity and sound velocity. Thus, the effect of higher temperatures and vibrational heat capacity acts to increase the difference between sound speed in monatomic vs. polyatomic molecules, with the speed remaining greater in monatomics.

Practical application to air

By far the most important factor influencing the speed of sound in air is temperature. The speed is proportional to the square root of the absolute temperature, giving an increase of about 0. 6 m/s per degree Celsius. For this reason, the pitch of a musical wind instrument increases as its temperature increases.

The speed of sound is raised by humidity but decreased by carbon dioxide. The difference between 0% and 100% humidity is about 1. 5 m/s at standard pressure and temperature, but the size of the humidity effect increases dramatically with temperature. The carbon dioxide content of air is not fixed, due to both carbon pollution and human breath (e. g., in the air blown through wind instruments).

The dependence on frequency and pressure are normally insignificant in practical applications. In dry air, the speed of sound increases by about 0. 1 m/s as the frequency rises from 10 Hz to 100 Hz. For audible frequencies above 100 Hz it is relatively constant. Standard values of the speed of sound are quoted in the limit of low frequencies, where the wavelength is large compared to the mean free path.

Mach number

Mach number, a useful quantity in aerodynamics, is the ratio of air speed to the local speed of sound. At altitude, for reasons explained, Mach number is a function of temperature. Aircraft flight instruments, however, operate using pressure differential to compute Mach number, not temperature. The assumption is that a particular pressure represents a particular altitude and, therefore, a standard temperature. Aircraft flight instruments need to operate this way because the stagnation pressure sensed by a Pitot tube is dependent on altitude as well as speed.

Experimental methods

A range of different methods exist for the measurement of sound in air.

The earliest reasonably accurate estimate of the speed of sound in air was made by William Derham, and acknowledged by Isaac Newton. Derham had a telescope at the top of the tower of the Church of St Laurence in Upminster, England. On a calm day, a synchronized pocket watch would be given to an assistant who would fire a shotgun at a pre-determined time from a conspicuous point some miles away, across the countryside. This could be confirmed by telescope. He then measured the interval between seeing gunsmoke and arrival of the sound using a half-second pendulum. The distance from where the gun was fired was found by triangulation, and simple division (distance / time) provided velocity. Lastly, by making many observations, using a range of different distances, the inaccuracy of the half-second pendulum could be averaged out, giving his final estimate of the speed of sound. Modern stopwatches enable this method to be used today over distances as short as 200–400 meters, and not needing something as loud as a shotgun.

Single-shot timing methods

The simplest concept is the measurement made using two microphones and a fast recording device such as a digital storage scope. This method uses the following idea.

If a sound source and two microphones are arranged in a straight line, with the sound source at one end, then the following can be measured:

1. The distance between the microphones (x), called microphone basis.

2. The time of arrival between the signals (delay) reaching the different microphones (t).

Then v = x / t.

Other methods

In these methods the time measurement has been replaced by a measurement of the inverse of time (frequency).

Kundt's tube is an example of an experiment which can be used to measure the speed of sound in a small volume. It has the advantage of being able to measure the speed of sound in any gas. This method uses a powder to make the nodes and antinodes visible to the human eye. This is an example of a compact experimental setup.

A tuning fork can be held near the mouth of a long pipe which is dipping into a barrel of water. In this system it is the case that the pipe can be brought to resonance if the length of the air column in the pipe is equal to ({1+2n}ë/4) where n is an integer. As the antinodal point for the pipe at the open end is slightly outside the mouth of the pipe it is best to find two or more points of resonance and then measure half a wavelength between these.

Here it is the case that v = fë.

High-precision measurements in air

The effect from impurities can be significant when making high-precision measurements. Chemical dessicants can be used to dry the air, but will in turn contaminate the sample. The air can be dried cryogenically, but this has the effect of removing the carbon dioxide as well; therefore many high-precision measurements are performed with air free of carbon dioxide rather than with natural air. A 2002 review found that a 1963 measurement by Smith and Harlow using a cylindrical resonator gave "the most probable value of the standard sound speed to date. " The experiment was done with air from which the carbon dioxide had been removed, but the result was then corrected for this effect so as to be applicable to real air. The experiments were done at 30°C but corrected for temperature in order to report them at 0°C. The result was 331. 45±0. 01 m/s for dry air at STP, for frequencies from 93 Hz to 1500 Hz.

NON-GASEOUS MEDIA

Speed of sound in solids

In a solid, there is a non-zero stiffness both for volumetric deformations and shear deformations. Hence, it is possible to generate sound waves with different velocities dependent on the deformation mode. Sound waves generating volumetric deformations (compression) and shear deformations (shearing) are called pressure waves (longitudinal waves) and shear waves (transverse waves), respectively. In earthquakes, the corresponding seismic waves are called P-waves (primary waves) and S-waves (secondary waves), respectively. The sound velocities of these two types of waves propagating in a homogeneous 3-dimensional solid are respectively given bywhere

- K is the bulk modulus of the elastic materials;
- G is the shear modulus of the elastic materials;
- E is the Young's modulus;
- ρ is the density;
- ν is Poisson's ratio.

The last quantity is not an independent one, as $E = 3K(1-2\nu)$. Note that the speed of pressure waves depends both on the pressure and shear resistance properties of the material, while the speed of shear waves depends on the shear properties only.

Typically, pressure waves travel faster in materials than do shear waves, and in earthquakes this is the reason that the onset of an earthquake is often preceded by a quick upward-downward shock, before arrival of waves that produce a side-to-side motion. For example, for a typical steel alloy, K = 170 GPa, G = 80 GPa and = 7700 kg/m, yielding a compressional speed $c_{solid'}$ p of 6000 m/s. This is in reasonable agreement with $c_{solid'}$ p measured experimentally at 5930 m/s for a (possibly different) type of steel. The shear speed $c_{solid',\ s}$ is estimated at 3200 m/s using the same numbers.

One-dimensional solids

The speed of sound for pressure waves in stiff materials such as metals is sometimes given for "long rods" of the material in question, in which the speed is easier to measure. In rods where their diameter is shorter than a wavelength, the speed of pure pressure waves may be simplified and is given by:

$$c_{\mathrm{solid}} = \sqrt{\frac{E}{\rho}},$$

where E is the Young's modulus. This is similar to the expression for shear waves, save that Young's modulus replaces the shear modulus. This speed of sound for pressure waves in long rods will always be slightly less than the same speed in homogeneous 3-dimensional solids, and the ratio of the speeds in the two different types of objects depends onPoisson's ratio for the material.

Speed of sound in liquids

In a fluid the only non-zero stiffness is to volumetric deformation (a fluid does not sustain shear forces).

Hence the speed of sound in a fluid is given by

$$c_{\mathrm{fluid}} = \sqrt{\frac{K}{\rho}},$$

where K is the bulk modulus of the fluid.

Water

In fresh water, sound travels at about 1497 m/s at 25 °C. See Technical Guides - Speed of Sound in Pure Water for an online calculator. Applications of underwater sound can be found in sonar, acoustic communication and acoustical oceanography. See Discovery of Sound in the Sea for other examples of the uses of sound in the ocean (by both man and other animals).

Seawater

In salt water that is free of air bubbles or suspended sediment, sound travels at about 1500 m/s. The speed of sound in seawater depends on pressure (hence depth), temperature (a change of 1 °C ~ 4 m/s), and salinity (a change of 1‰ ~ 1 m/s), and empirical equations have been derived to accurately calculate sound speed from these variables. Other factors affecting sound speed are minor. Since temperature decreases with depth while pressure and generally salinity increase, the profile of sound speed with depth generally shows a characteristic curve which decreases to a minimum at a depth of several hundred meters, then increases again with increasing depth (right). For more information see Dushaw et al.

A simple empirical equation for the speed of sound in sea water with reasonable accuracy for the world's oceans is due to Mackenzie:

$$c(T,S,z) = a_1 + a_2T + a_3T^2 + a_4T^3 + a_5(S-35) + a_6z + a_7z^2 + a_8T(S-35) + a_9Tz^3,$$

where T, S, and z are temperature in degrees Celsius, salinity in parts per thousand and depth in meters, respectively. The constants a1, a2, …, a9 are:

$$a_1 = 1448.96, \quad a_2 = 4.591, \quad a_3 = -5.304 \times 10^{-2},$$
$$a_4 = 2.374 \times 10^{-4}, \quad a_5 = 1.340, \quad a_6 = 1.630 \times 10^{-2},$$
$$a_7 = 1.675 \times 10^{-7}, \quad a_8 = -1.025 \times 10^{-2}, \quad a_9 = -7.139 \times 10^{-13},$$

with check value 1550. 744 m/s for T = 25 °C, S = 35 parts per thousand, z = 1000 m. This equation has a standard error of 0. 070 m/s for salinity between 25 and 40 ppt. See Technical Guides. Speed of Sound in Sea-Water for an online calculator.

Other equations for sound speed in sea water are accurate over a wide range of conditions, but are far more complicated, e. g., that by V. A. Del Grosso and the Chen-Millero-Li Equation.

Speed in plasma

The speed of sound in a plasma for the common case that the electrons are hotter than the ions (but not too much hotter) is given by the formula

$$c_s = (\gamma Z k T_e / m_i)^{1/2} = 9.79 \times 10^3 (\gamma Z T_e / \mu)^{1/2} \text{ m/s},$$

where

- m_i is the ion mass;
- μ is the ratio of ion mass to proton mass $\mu = m_i/m_p$;
- T_e is the electron temperature;
- Z is the charge state;
- k is Boltzmann's constant;
- γ is the adiabatic index.

In contrast to a gas, the pressure and the density are provided by separate species, the pressure by the electrons and the density by the ions. The two are coupled through a fluctuating electric field.

Gradients

When sound spreads out evenly in all directions in three dimensions, the intensity drops in proportion to the inverse square of the distance. However, in the ocean there is a layer called the 'deep sound channel' or SOFAR channel which can confine sound waves at a particular depth.

In the SOFAR channel, the speed of sound is lower than that in the layers above and below. Just as light waves will refract towards a region of higher index, sound waves willrefract towards a region where their speed is reduced. The result is that sound gets confined in the layer, much the way light can be confined in a sheet of glass or optical fiber. Thus, the sound is confined in essentially two dimensions. In two dimensions the intensity drops in proportion to only the inverse of the distance. This allows waves to travel much further before being undetectably faint.

A similar effect occurs in the atmosphere. Project Mogul successfully used this effect to detect a nuclear explosion at a considerable distance.

Sound pressure

Sound pressure or acoustic pressure is the local pressure deviation from the ambient (average, or equilibrium) atmospheric pressure, caused by a sound wave. In air, sound pressure can be measured using a microphone, and in water with a hydrophone. The SI unit of sound pressure is the pascal (Pa). Mathematical definition

A sound wave in a transmission medium causes a deviation (sound pressure, a dynamic pressure) in the local ambient pressure, a staticpressure.

Sound pressure, denoted p, is defined by $p_{total} = p_{stat} + p$,

where

- p_{total} is the total pressure;
- p_{stat} is the static pressure.

Sound measurements

Sound intensity

In a sound wave, the complementary variable to sound pressure is the particle velocity. Together they determine the sound intensityof the wave.

Sound intensity, denoted I and measured in W·m in SI units, is defined by I = pv.

where

- p is the sound pressure;
- v is the particle velocity.

Acoustic impedance

Acoustic impedance, denoted Z and measured in Pa·m·s in SI units, is defined by

$$Z(s) = \frac{\hat{p}(s)}{\hat{Q}(s)},$$

where

- p̂(Q) is the Laplace transform of sound pressure;
- Q̂(s) is the Laplace transform of sound volume flow rate.

Specific acoustic impedance, denoted z and measured in Pa·m·s in SI units, is defined by

$$z(s) = \frac{\hat{p}(s)}{\hat{v}(s)},$$

where

- p̂is the Laplace transform of sound pressure;
- v̂is the Laplace transform of particle velocity.

Particle displacement

The particle displacement of a progressive sine wave is given by

$$\xi(\mathbf{r}, t) = \xi_\mathrm{m} \cos(\mathbf{k} \cdot \mathbf{r} - \omega t + \varphi_{\xi,0}),$$

where

- î$_m$ is the amplitude of the particle displacement;
- φξ,0 is the phase shift of the particle displacement;
- k is the angular wavevector;
- ù is the angular frequency.

It follows that the particle velocity and the sound pressure along the direction of propagation of the sound wave x are given by

$$v(\mathbf{r}, t) = \frac{\partial \xi}{\partial t}(\mathbf{r}, t) = \omega \xi_\mathrm{m} \cos\left(\mathbf{k} \cdot \mathbf{r} - \omega t + \varphi_{\xi,0} + \frac{\pi}{2}\right) = v_\mathrm{m} \cos(\mathbf{k} \cdot \mathbf{r} - \omega t + \varphi_{v,0}),$$

where

- v_m is the amplitude of the particle velocity;
- ϕυ,0 is the phase shift of the particle velocity;
- p_m is the amplitude of the acoustic pressure;
- ϕp,0 is the phase shift of the acoustic pressure.

Inverse-proportional law

When measuring the sound pressure created by an object, it is important to measure the distance from the object as well, since the sound pressure of a spherical sound wave decreases as 1/r from the centre of the sphere (and not as 1/r, like the sound intensity):

$$p(r) \propto \frac{1}{r}.$$

This relationship is an inverse-proportional law.

If the sound pressure p_1 is measured at a distance r1 from the centre of the sphere, the sound pressure p_2 at another position r_2 can be calculated:

$$p_2 = \frac{r_1}{r_2}\, p_1.$$

The inverse-proportional law for sound pressure comes from the inverse-square law for sound intensity:

$$I(r) \propto \frac{1}{r^2}.$$

Indeed, $I(r) = p(r)v(r) = p(r)[p * z^{-1}](r) \propto p^2(r),$

where

- * is the convolution operator;
- z is the convolution inverse of the specific acoustic impedance,

hence the inverse-proportional law:

The sound pressure may vary in direction from the centre of the sphere as well, so measurements at different angles may be necessary, depending on the situation. An obvious example of a sound source whose spherical sound wave varies in level in different directions is a bullhorn.

Sound pressure level

Sound pressure level (SPL) or acoustic pressure level is a logarithmic measure of the effective pressure of a sound relative to a reference value.

Sound pressure level, denoted L_P $L_p = \ln\left(\frac{p}{p_0}\right)$ Np $= 2\log_{10}\left(\frac{p}{p_0}\right)$ B $= 20\log_{10}\left(\frac{p}{p_0}\right)$ dB, and measured in dB, is defined bywhere

- p is the root mean square sound pressure;
- p_0 is the reference sound pressure;
- 1 Np = 1 is the neper;
- 1 B = (1/2) ln(10) is the bel;
- 1 dB = (1/20) ln(10) is the decibel.

The commonly used reference sound pressure in air is p_0 = 20 μPa.

which is often considered as the threshold of human hearing (roughly the sound of a mosquito flying 3 m away). The proper notations for sound pressure level using this reference are $L_{p/(20\ \text{ìPa})}$ or L_p (re 20 ìPa), but the suffix notations dB SPL, dB(SPL), dBSPL, or dB_{SPL} are very common, even if they are not accepted by the SI.

Most sound level measurements will be made relative to this reference, meaning 1 Pa will equal an SPL of 94 dB. In other media, such as underwater, a reference level of 1 ìPa is used. These references are defined in ANSI S1. 1-1994.

Examples

The lower limit of audibility is defined as SPL of 0 dB, but the upper limit is not as clearly defined. While 1 atm (194 dB Peak or 191 dB SPL) is the largest pressure variation an undistorted sound wave can have in Earth's atmosphere, larger sound waves can be present in other atmospheres or other media such as under water, or through the Earth.

Ears detect changes in sound pressure. Human hearing does not have a flat spectral sensitivity (frequency response) relative to frequency versus amplitude. Humans do not perceive low- and high-frequency sounds as well as they perceive sounds near2, 000 Hz, as shown in the equal-loudness contour. Because the frequency response of human hearing changes with amplitude, three weightings have been established for measuring sound pressure: A, B and C. A-weighting applies to sound pressures levels up to 55 dB, B-weighting applies to sound pressures levels between 55 dB and 85 dB, and C-weighting is for measuring sound pressure levels above 85 dB.

In order to distinguish the different sound measures a suffix is used: A-weighted sound pressure level is written either as dB_A or L_A. B-weighted sound pressure level is written either as dB_B or L_B, and C-weighted sound pressure level is written either as dBC or LC. Unweighted sound pressure level is called "linear sound pressure level" and is often written as dB_L or just L. Some sound measuring instruments use the letter "Z" as an indication of linear SPL.

Distance

The distance of the measuring microphone from a sound source is often omitted when SPL measurements are quoted, making the data useless. In the case of ambient environmental measurements of "background" noise, distance need not be quoted as no single source is present, but when measuring the noise level of a specific piece of equipment the distance should always be stated. A distance of one metre (1 m) from the source is a frequently used standard distance. Because of the effects of reflected noise within a closed room, the use of an anechoic chamber allows for sound to be comparable to measurements made in a free field environment.

According to the inverse proportional law, when sound level Lp1 is measured at a distance r_1, the sound level L_{p2} at the distance r_2 is

$$L_{p2} = L_{p1} + 20 \log_{10}\left(\frac{r_1}{r_2}\right) \text{ dB}.$$

9

ELECTRIC CURRENT

An electric current is a flow of electric charge. In electric circuits this charge is often carried by moving electrons in a wire. It can also be carried by ions in an electrolyte, or by both ions and electrons such as in a plasma.

The SI unit for measuring an electric current is the ampere, which is the flow of electric charge across a surface at the rate of onecoulomb per second. Electric current is measured using a device called an ammeter.

Electric currents cause Joule heating, which creates light in incandescent light bulbs. They also create magnetic fields, which are used in motors, inductors and generators.

The particles that carry the charge in an electric current are called charge carriers. In metals, one or more electrons from each atom are loosely bound to the atom, and can move freely about within the metal. These conduction electrons are the charge carriers in metal conductors.

SYMBOL

The conventional symbol for current is I, which originates from the French phrase intensité de courant, or in English current intensity. This phrase is frequently used when discussing the value of an electric current, but modern practice often shortens this to simply current.

The I symbol was used by André-Marie Ampère, after whom the unit of electric current is named, in formulating the eponymous Ampère's force law, which he discovered in 1820. The notation travelled from France to Great Britain, where it became standard, although at least one journal did not change from using C to I until 1896.

Conventions

In metals, which make up the wires and other conductors in most electrical circuits, the positively charged atomic nuclei are held in a fixed position, and the electrons are free to move, carrying their charge from one place to another. In other materials, notably thesemiconductors, the charge carriers can be positive or negative, depending on the dopant used. Positive and negative charge carriers may even be present at the same time, as happens in an electrochemical cell.

A flow of positive charges gives the same electric current, and has the same effect in a circuit, as an equal flow of negative charges in the opposite direction. Since current can be the flow of either positive or negative charges, or both, a convention is needed for the direction of current that is independent of the type of charge carriers. The direction of conventional current is arbitrarily defined as the same direction as positive charges flow.

The consequence of this convention is that electrons, the charge carriers in metal wires and most other parts of electric circuits, flow in the opposite direction of conventional current flow in an electrical circuit.

Reference direction

Since the current in a wire or component can flow in either direction, when a variable I is defined to represent that current, the direction representing positive current must be specified, usually by an arrow on the circuit schematic diagram. This is called the reference direction of current I. If the current flows in the opposite direction, the variable I has a negative value.

When analyzing electrical circuits, the actual direction of current through a specific circuit element is usually unknown. Consequently, the reference directions of currents are often assigned arbitrarily. When the circuit is solved, a negative value for the variable means that the actual direction of current through that circuit element is opposite that of the chosen reference direction. In electronic circuits, the reference current directions are often chosen so that all currents are toward ground. This often corresponds to the actual current direction, because in many circuits the power supply voltage is positive with respect to ground.

Ohm's law

Ohm's law states that the current through a conductor between two points is directly proportional to the potential difference across the two points. Introducing the constant of proportionality, the resistance, one arrives at the usual mathematical equation that describes this relationship: $I = \frac{V}{R}$

where I is the current through the conductor in units of amperes, V is the potential difference measured across the conductor in units of volts, and R is the resistance of the conductor in units of ohms. More specifically, Ohm's law states that the R in this relation is constant, independent of the current.

AC and DC

The abbreviations AC and DC are often used to mean simply alternating and direct, as when they modify current or voltage.

Direct current

Direct current (DC) is the unidirectional flow of electric charge. Direct current is produced by sources such as batteries, thermocouples, solar cells, and commutator-type electric machines of the dynamo type. Direct current may flow in a conductor such as a wire, but can also flow through semiconductors, insulators, or even through a vacuum as in electron or ion beams. The electric charge flows in a constant direction, distinguishing it from alternating current (AC). A term formerly used for direct current was galvanic current. Direct current (DC) is the unidirectional flow of electric charge. Direct current is produced by sources such as batteries, thermocouples, solar cells, and commutator-type electric machines of the dynamo type. Direct current may flow in a conductor such as a wire, but can also flow through semiconductors, insulators, or even through a vacuum as in electron or ion beams. The electric current flows in a constant direction, distinguishing it from alternating current (AC). A term formerly used for this type of current was galvanic current.

The abbreviations AC and DC are often used to mean simply alternating and direct, as when they modify current or voltage.

Direct current may be obtained from an alternating current supply by use of a current-switching arrangement called a rectifier, which contains electronic elements (usually) or electromechanical elements (historically) that allow current to flow only in one direction. Direct current may be made into alternating current with an inverter or a motor-generator set.

The first commercial electric power transmission (developed by Thomas Edison in the late nineteenth century) used direct current. Because of the significant advantages of alternating current over direct current in transforming and transmission, electric power distribution is nearly all alternating current today. In the mid-1950s, high-voltage direct current transmission was developed, and is now an option instead of long-distance high voltage alternating current systems. For long distance underseas cables (e. g. between countries, such as NorNed), this

DC option is the only technically feasible option. For applications requiring direct current, such as third rail power systems, alternating current is distributed to a substation, which utilizes a rectifier to convert the power to direct current. See War of Currents.

Direct current is used to charge batteries, and in nearly all electronic systems, as the power supply. Very large quantities of direct-current power are used in production of aluminumand other electrochemical processes. Direct current is used for some railway propulsion, especially in urban areas. High-voltage direct current is used to transmit large amounts of power from remote generation sites or to interconnect alternating current power grids.

Various definitions

The term DC is used to refer to power systems that use only one polarity of voltage or current, and to refer to the constant, zero-frequency, or slowly varying local mean value of a voltage or current. For example, the voltage across a DC voltage source is constant as is the current through a DC current source. The DC solution of an electric circuit is the solution where all voltages and currents are constant. It can be shown that any stationary voltage or current waveform can be decomposed into a sum of a DC component and a zero-mean time-varying component; the DC component is defined to be the expected value, or the average value of the voltage or current over all time.

Although DC stands for "direct current", DC often refers to "constant polarity". Under this definition, DC voltages can vary in time, as seen in the raw output of a rectifier or the fluctuating voice signal on a telephone line.

Some forms of DC (such as that produced by a voltage regulator) have almost no variations in voltage, but may still have variations in output power and current.

Circuits

A direct current circuit is an electrical circuit that consists of any combination of constant voltage sources, constant current sources, andresistors. In this case, the circuit voltages and currents are independent of time. A particular circuit voltage or current does not depend on the past value of any circuit voltage or current. This implies that the system of equations that represent a DC circuit do not involve integrals or derivatives with respect to time.

If a capacitor or inductor is added to a DC circuit, the resulting circuit is not, strictly speaking, a DC circuit. However, most such circuits have a DC solution. This solution gives the circuit voltages and currents when the circuit is in DC steady state. Such a circuit is represented by a system of differential equations. The solution to these equations usually contain a time varying or transient part as well as constant or steady state part. It is this steady state part that is the DC solution. There are some circuits that do not have a DC solution. Two simple examples are a constant current source connected to a capacitor and a constant voltage source connected to an inductor.

In electronics, it is common to refer to a circuit that is powered by a DC voltage source such as a battery or the output of a DC power supply as a DC circuit even though what is meant is that the circuit is DC powered.

Applications

Direct-current installations usually have different types of sockets, connectors, switches, and fixtures, mostly due to the low voltages used, from those suitable for alternating current. It is usually important with a direct-current appliance not to reverse polarity unless the device has a diode bridge to correct for this (most battery-powered devices do not).

The Unicode code point for the direct current symbol, found in the Miscellaneous Technical block, is U+2393 ("#).

DC is commonly found in many extra-low voltage applications and some low-voltage applications, especially where these are powered bybatteries, which can produce only DC, or solar power systems, since solar cells can produce only DC. Most automotive applications use DC, although the alternator is an AC device which uses a rectifier to produce DC. Most electronic circuits require a DC power supply. Applications using fuel cells (mixing hydrogen and oxygen together with a catalyst to produce electricity and water as byproducts) also produce only DC.

The vast majority of automotive applications use "12-volt" DC power; a few have a 6 V or a 42 V electrical system.

Light aircraft electrical systems are typically 12 V or 28 V.

Through the use of a DC-DC converter, high DC voltages such as 48 V to 72 V DC can be stepped down to 36 V, 24 V, 18 V, 12 V or 5 V to supply different loads. In a telecommunications system operating at 48 V DC, it is generally more efficient to step voltage down to 12 V to 24 V DC with a DC-DC converter and power equipment loads directly at their native DC input voltages versus operating a 48 V DC to 120 V AC inverter to provide power to equipment.

Many telephones connect to a twisted pair of wires, and use a bias tee to internally separate the AC component of the voltage between the two wires (the audio signal) from the DC component of the voltage between the two wires (used to power the phone).

Telephone exchange communication equipment, such as DSLAM, uses standard "48 V DC power supply. The negative polarity is achieved by grounding the positive terminal of power supply system and the battery bank. This is done to prevent electrolysis depositions.

Alternating current

In alternating current (AC, also ac), the movement of electric charge periodically reverses direction. In direct current (DC, also dc), the flow of electric charge is only in one direction.

AC is the form of electric power delivered to businesses and residences. The usual waveform of an AC power circuit is a sine wave. Certain applications use different waveforms, such as triangular or square waves. Audio and radio signals carried on electrical wires are also examples of alternating current. An important goal in these applications is recovery of information encoded (or modulated) onto the AC signal.

Transmission, distribution, and domestic power supply

AC voltage may be increased or decreased with a transformer. Use of a higher voltage leads to significantly more efficient transmission of power. The power losses (P_L) in a conductor are a product of the square of the current (I) and the resistance (R) of the conductor, described by the formula $P_L = I_2 R$.

This means that when transmitting a fixed power on a given wire, if the current is doubled, the power loss will be four times greater.

The power transmitted is equal to the product of the current and the voltage (assuming no phase difference); that is, $P_T = IV$.

Thus, the same amount of power can be transmitted with a lower current by increasing the voltage. It is therefore advantageous when transmitting large amounts of power to distribute the power with high voltages (often hundreds of kilovolts).

However, high voltages also have disadvantages, the main one being the increased insulation required, and generally increased difficulty in their safe handling. In a power plant, power is generated at a convenient voltage for the design of a generator, and then stepped up to a high voltage for transmission. Near the loads, the transmission voltage is stepped down to the voltages used by equipment. Consumer voltages vary depending on the country and size of load, but generally motors and lighting are built to use up to a few hundred volts between phases.

The utilization voltage delivered to equipment such as lighting and motor loads is standardized, with an allowable range of voltage over which equipment is expected to operate. Standard power utilization voltages and percentage tolerance vary in the different mains power systems found in the world.

Modern high-voltage direct-current (HVDC) electric power transmission systems contrast with the more common alternating-current systems as a means for the efficient bulk transmission of electrical power over long distances. HVDC systems, however, tend to be more expensive and less efficient over shorter distances than transformers. Transmission with high voltage direct current was not feasible when Edison, Westinghouse and Tesla were designing their power systems, since there was then no way to economically convert AC power to DC and back again at the necessary voltages.

Three-phase electrical generation is very common. The simplest case is three separate coils in the generator stator that are physically offset by an angle of 120° to each other. Three current waveforms are produced that are equal in magnitude and 120° out of phase to each other. If coils are added opposite to these (60° spacing), they generate the same phases with reverse polarity and so can be simply wired together.

In practice, higher "pole orders" are commonly used. For example, a 12-pole machine would have 36 coils (10° spacing). The advantage is that lower speeds can be used. For example, a 2-pole machine running at 3600 rpm and a 12-pole machine running at 600 rpm produce the same frequency. This is much more practical for larger machines.

If the load on a three-phase system is balanced equally among the phases, no current flows through the neutral point. Even in the worst-case unbalanced (linear) load, the neutral current will not exceed the highest of the phase currents. Non-linear loads (e. g., computers) may require an oversized neutral bus and neutral conductor in the upstream distribution panel to handle harmonics. Harmonics can cause neutral conductor current levels to exceed that of one or all phase conductors.

For three-phase at utilization voltages a four-wire system is often used. When stepping down three-phase, a transformer with a Delta (3-wire) primary and a Star (4-wire, center-earthed) secondary is often used so there is no need for a neutral on the supply side.

For smaller customers (just how small varies by country and age of the installation) only a single phase and the neutral or two phases and the neutral are taken to the property. For larger installations all three phases and the neutral are taken to the main distribution panel. From the three-phase main panel, both single and three-phase circuits may lead off.

Three-wire single-phase systems, with a single center-tapped transformer giving two live conductors, is a common distribution scheme for residential and small commercial buildings in North America. This arrangement is sometimes incorrectly referred to as "two phase". A similar method is used for a different reason on construction sites in the UK. Small power tools and lighting are supposed to be supplied by a local center-tapped transformer with a voltage of 55 V between each power conductor and earth. This significantly reduces the risk of electric shock in the event that one of the live conductors becomes exposed through an equipment fault whilst still allowing a reasonable voltage of 110 V between the two conductors for running the tools.

A third wire, called the bond (or earth) wire, is often connected between non-current-carrying metal enclosures and earth ground. This conductor provides protection from electric shock due to accidental contact of circuit conductors with the metal chassis of portable appliances and tools. Bonding all non-current-carrying metal parts into one complete system ensures there is always a low electrical impedance path to ground sufficient to carry any fault current for as long as it takes for the system to clear the fault. This low impedance path allows the maximum amount of fault current, causing the overcurrent protection device (breakers, fuses) to trip or burn out as quickly as possible, bringing the electrical system to a safe state. All bond wires are bonded to ground at the main service panel, as is the Neutral/Identified conductor if present.

OCCURRENCES

Natural observable examples of electrical current include lightning, static electricity, and the solar wind, the source of the polar auroras.

Man-made occurrences of electric current include the flow of conduction electrons in metal wires such as the overhead power lines that deliver electrical energy across long distances and the smaller wires within electrical and electronic equipment. Eddy currents are electric currents that occur in conductors exposed to changing magnetic fields. Similarly, electric currents occur, particularly in the surface, of conductors exposed to electromagnetic waves. When oscillating electric currents flow at the correct voltages within radio antennas, radio waves are generated.

In electronics, other forms of electric current include the flow of electrons through resistors or through the vacuum in a vacuum tube, the flow of ions inside a battery or a neuron, and the flow of holes within a semiconductor.

Current measurement

Current can be measured using an ammeter.

At the circuit level, there are various techniques that can be used to measure current:

- Shunt resistors
- Hall effect current sensor transducers
- Transformers (however DC cannot be measured)
- Magnetoresistive field sensors

Resistive heating

Joule heating, also known as ohmic heating and resistive heating, is the process by which the passage of an electric current through a conductor releases heat. It was first studied by James Prescott Joule in 1841. Joule immersed a length of wire in a fixed mass of water and measured the temperature rise due to a known current through the wire for a 30minute period. By varying the current and the length of the wire he deduced that the heat produced was proportional to the square of the current multiplied by the electrical resistance of the wire. $Q \propto I^2 R$

This relationship is known as Joule's First Law. The SI unit of energy was subsequently named the joule and given the symbol J. The commonly known unit of power, the watt, is equivalent to one joule per second.

ELECTROMAGNETISM

Electromagnet

Electric current produces a magnetic field. The magnetic field can be visualized as a pattern of circular field lines surrounding the wire that persists as long as there is current.

Magnetism can also produce electric currents. When a changing magnetic field is applied to a conductor, an Electromotive force (EMF) is produced, and when there is a suitable path, this causes current.

Electric current can be directly measured with a galvanometer, but this method involves breaking the electrical circuit, which is sometimes inconvenient. Current can also be measured without breaking the circuit by detecting the magnetic field associated with the current. Devices used for this include Hall effect sensors, current clamps, current transformers, and Rogowski coils.

Radio waves

When an electric current flows in a suitably shaped conductor at radio frequencies radio waves can be generated. These travel at the speed of light and can cause electric currents in distant conductors.

Conduction mechanisms in various media

In metallic solids, electric charge flows by means of electrons, from lower to higher electrical potential. In other media, any stream of charged objects (ions, for example) may constitute an electric current. To provide a definition of current independent of the type of charge carriers, conventional current is defined as moving in the same direction as the positive charge flow. So, in metals where the charge carriers (electrons) are negative, conventional current is in the opposite direction as the electrons. In conductors where the charge carriers are positive, conventional current is in the same direction as the charge carriers.

In a vacuum, a beam of ions or electrons may be formed. In other conductive materials, the electric current is due to the flow of both positively and negatively charged particles at the same time. In still others, the current is entirely due to positive charge flow. For example, the electric currents in electrolytes are flows of positively and negatively charged ions. In a common lead-acid electrochemical cell, electric currents are composed of positive hydrogen ions (protons) flowing in one direction, and negative sulfate ions flowing in the other. Electric currents in sparks or plasma are flows of electrons as well as positive and negative ions. In ice and in certain solid electrolytes, the electric current is entirely composed of flowing ions.

Metals

A solid conductive metal contains mobile, or free electrons, which function as conduction electrons. These electrons are bound to the metal lattice but no longer to an individual atom. Metals are particularly conductive because there are a large number of these free electrons, typically one per atom in the lattice. Even with no external electric field applied, these electrons move about randomly due to thermal energy but, on average, there is zero net current within the metal. At room temperature, the average speed of these random motions is 10 metres per second. Given a surface through which a metal wire passes, electrons move in both directions across the surface at an equal rate. As George Gamowwrote in his popular science book, One, Two, Three... Infinity (1947), "The metallic substances differ from all other materials by the fact that the outer shells of their atoms are bound rather loosely, and often let one of their electrons go free. Thus the interior of a metal is filled up with a large number of unattached electrons that travel aimlessly around like a crowd of displaced persons. When a metal wire is subjected to electric force applied on its opposite ends, these free electrons rush in the direction of the force, thus forming what we call an electric current. "

When a metal wire is connected across the two terminals of a DC voltage source such as a battery, the source places an electric field across the conductor. The moment contact is made, the free electrons of the conductor are forced to drift toward the positive terminal under the influence of this field. The free electrons are therefore the charge carrier in a typical solid conductor.

For a steady flow of charge through a surface, the current I (in amperes) can be calculated with the following equation: $I = \frac{Q}{t}$,

where Q is the electric charge transferred through the surface over a time t. If Q and t are measured in coulombs and seconds respectively, I is in amperes.

More generally, electric current can be represented as the rate at which charge flows through a given surface as:

$$I = \frac{dQ}{dt}.$$

Electrolytes

Electric currents in electrolytes are flows of electrically charged particles (ions). For example, if an electric field is placed across a solution of Na and Cl (and conditions are right) the sodium ions move towards the negative electrode (cathode), while the chloride ions move towards the positive electrode (anode). Reactions take place at both electrode surfaces, absorbing each ion.

Water-ice and certain solid electrolytes called proton conductors contain positive hydrogen ions ("protons") that are mobile. In these materials, electric currents are composed of moving protons, as opposed to the moving electrons in metals.

In certain electrolyte mixtures, brightly coloured ions are the moving electric charges. The slow progress of the colour makes the current visible.

Gases and plasmas

In air and other ordinary gases below the breakdown field, the dominant source of electrical conduction is via relatively few mobile ions produced by radioactive gases, ultraviolet light, or cosmic rays. Since the electrical conductivity is low, gases are dielectrics or insulators. However, once the applied electric field approaches the breakdown value, free electrons become sufficiently accelerated by the electric field to create additional free electrons by colliding, and ionizing, neutral gas atoms or molecules in a process calledavalanche breakdown. The breakdown process forms a plasma that contains enough mobile electrons and positive ions to make it an electrical conductor. In the process, it forms a light emitting conductive path, such as a spark, arc or lightning.

Plasma is the state of matter where some of the electrons in a gas are stripped or "ionized" from their molecules or atoms. A plasma can be formed by high temperature, or by application of a high electric or alternating magnetic field as noted above. Due to their lower mass, the electrons in a plasma accelerate more quickly in response to an electric field than the heavier positive ions, and hence carry the bulk of the current. The free ions recombine to create new chemical compounds (for example, breaking atmospheric oxygen into single oxygen [O2 '! 2O], which then recombine creating ozone [O3]).

Vacuum

Since a "perfect vacuum" contains no charged particles, it normally behaves as a perfect insulator. However, metal electrode surfaces can cause a region of the vacuum to become conductive by injecting free electrons or ions through either field electron emission or thermionic emission. Thermionic emission occurs when the thermal energy exceeds the metal'swork function, while field electron emission occurs when the electric field at the surface of the metal is high enough to cause tunneling, which results in the ejection of free electrons from the metal into the vacuum. Externally heated electrodes are often used to generate an electron cloud as in the filament or indirectly heated cathode of vacuum tubes. Cold electrodes can also spontaneously produce electron clouds via thermionic emission when small incandescent regions (called cathode spots or anode spots) are formed. These are incandescent regions of the electrode surface that are created by a localized high current. These regions may be initiated by field electron emission, but are then sustained by localized thermionic emission once a vacuum arc forms. These small electron-emitting regions can form quite rapidly, even explosively, on a metal surface subjected to a high electrical field. Vacuum tubes and sprytrons are some of the electronic switching and amplifying devices based on vacuum conductivity.

Superconductivity

Superconductivity is a phenomenon of exactly zero electrical resistance and expulsion of magnetic fields occurring in certain materials when cooled below a characteristic critical temperature. It was discovered by Heike Kamerlingh Onnes on April 8, 1911 in Leiden. Like ferromagnetism and atomic spectral lines, superconductivity is a quantum mechanical phenomenon. It is characterized by the Meissner effect, the complete ejection of magnetic field lines from the interior of the superconductor as it transitions into the superconducting state. The occurrence of the Meissner effect indicates that superconductivity cannot be understood simply as the idealization of perfect conductivity in classical physics.

Semiconductor

In a semiconductor it is sometimes useful to think of the current as due to the flow of positive "holes" (the mobile positive charge carriers that are places where the semiconductor crystal is missing a valence electron). This is the case in a p-type semiconductor. A semiconductor has electrical conductivity intermediate in magnitude between that of a conductorand an insulator. This means a conductivity roughly in the range of 10 to 10 siemens per centimeter (SÅ"cm).

In the classic crystalline semiconductors, electrons can have energies only within certain bands (i. e. ranges of levels of energy). Energetically, these bands are located between the energy of the ground state, the state in which electrons are tightly bound to the atomic nuclei of the material, and the free electron energy, the latter describing the energy required for an electron to escape entirely from the material. The energy bands each correspond to a large number of discrete quantum states of the electrons, and most of the states with low energy (closer to the nucleus) are occupied, up to a particular band called the valence band. Semiconductors and insulators are distinguished from metals because the valence band in any given metal is nearly filled with electrons under usual operating conditions, while very few (semiconductor) or virtually none (insulator) of them are available in theconduction band, the band immediately above the valence band.

The ease of exciting electrons in the semiconductor from the valence band to the conduction band depends on the band gap between the bands. The size of this energy band gap serves as an arbitrary dividing line (roughly 4 eV) between semiconductors and insulators.

With covalent bonds, an electron moves by hopping to a neighboring bond. The Pauli exclusion principle requires that the electron be lifted into the higher anti-bonding state of that bond. For delocalized states, for example in one dimension – that is in a nanowire, for every energy there is a state with electrons flowing in one direction and another state with the electrons flowing in the other. For a net current to flow, more states for one direction than for the other direction must be occupied. For this to occur, energy is required, as in the semiconductor the next higher states lie above the band gap. Often this is stated as: full bands do not contribute to the electrical conductivity. However, as a semiconductor's temperature rises above absolute zero, there is more energy in the semiconductor to spend on lattice vibration and on exciting electrons into the conduction band. The current-carrying electrons in the conduction band are known as free electrons, though they are often simply called electrons if that is clear in context.

Current density and Ohm's law

Current density is a measure of the density of an electric current. It is defined as a vector whose magnitude is the electric current per cross-sectional area. In SI units, the current density is measured in amperes per square metre. $I = \int \vec{J} \cdot d\vec{A}$

where I is current in the conductor $\vec{J}$, is the current density, and $d\vec{A}$ is the differential cross-sectional area vector.

The current density (current per unit area) in materials with finite resistance is directly proportional to the electric field in the medium. The proportionality constant is called theconductivity of the material, whose value depends on the material concerned and, in general, is dependent on the temperature of the material: The reciprocal of the conductivity of the material is called the resistivity of the material and the above equation, when written in terms of resistivity becomes:

or

Conduction in semiconductor devices may occur by a combination of drift and diffusion, which is proportional to diffusion constant and charge density. The current density is then:

with being the elementary charge and the electron density. The carriers move in the direction of decreasing concentration, so for electrons a positive current results for a positive density gradient. If the carriers are holes, replace electron density by the negative of the hole density.

In linear anisotropic materials, ó, ñ and D are tensors.

In linear materials such as metals, and under low frequencies, the current density across the conductor surface is uniform. In such conditions, Ohm's law states that the current is directly proportional to the potential difference between two ends (across) of that metal (ideal) resistor (or other ohmic device)

where is the current, measured in amperes; is the potential difference, measured in volts; and is the resistance, measured in ohms. For alternating currents, especially at higher frequencies, skin effect causes the current to spread unevenly across the conductor cross-section, with higher density near the surface, thus increasing the apparent resistance.

DRIFT SPEED

The mobile charged particles within a conductor move constantly in random directions, like the particles of a gas. To create a net flow of charge, the particles must also move together with an average drift rate. Electrons are the charge carriers in metals and they follow an erratic path, bouncing from atom to atom, but generally drifting in the opposite direction of the electric field. The speed they drift at can be calculated from the equation: I = nAvQ.

where I

- is the electric current n
- is number of charged particles per unit volume (or charge carrier density) A
- is the cross-sectional area of the conductor v
- is the drift velocity, and Q
- is the charge on each particle.

Typically, electric charges in solids flow slowly. For example, in a copper wire of cross-section 0. 5 mm, carrying a current of 5 A, the drift velocity of the electrons is on the order of a millimetre per second. To take a different example, in the near-vacuum inside a cathode ray tube, the electrons travel in near-straight lines at about a tenth of the speed of light.

Any accelerating electric charge, and therefore any changing electric current, gives rise to an electromagnetic wave that propagates at very high speed outside the surface of the conductor. This speed is usually a significant fraction of the speed of light, as can be deduced from Maxwell's Equations, and is therefore many times faster than the drift velocity of the electrons. For example, in AC power lines, the waves of electromagnetic energy propagate through the space between the wires, moving from a source to a distant load, even though the electrons in the wires only move back and forth over a tiny distance.

The ratio of the speed of the electromagnetic wave to the speed of light in free space is called the velocity factor, and depends on the electromagnetic properties of the conductor and the insulating materials surrounding it, and on their shape and size.

The magnitudes (but, not the natures) of these three velocities can be illustrated by an analogy with the three similar velocities associated with gases.

- The low drift velocity of charge carriers is analogous to air motion; in other words, winds.
- The high speed of electromagnetic waves is roughly analogous to the speed of sound in a gas (these waves move through the medium much faster than any individual particles do)

- The random motion of charges is analogous to heat – the thermal velocity of randomly vibrating gas particles.

ELECTRICAL MEASUREMENTS

Electrical measurements are the methods, devices and calculations used to measure electrical quantities. Measurement of electrical quantities may be done to measure electrical parameters of a system. Using transducers, physical properties such as temperature, pressure, flow, force, and many others can be converted into electrical signals, which can then be conveniently measured and recorded. High-precision laboratory measurements of electrical quantities are used in experiments to determine fundamental physical properties such as the charge of the electron or the speed of light, and in the definition of the units for electrical measurements, with precision in some cases on the order of a few parts per million. Less precise measurements are required every day in industrial practice. Electrical measurements are a branch of the science of metrology.

Measurable independent and semi-independent electrical quantities comprise:

- Voltage
- Electric current
- Electrical resistance and electrical conductance
- Electrical reactance and susceptance
- Magnetic flux
- Electrical charge by the means of electrometer
- Magnetic field by the means of Hall sensor
- Electric field
- Electrical power by the means of electricity meter
- S-matrix by the means of network analyzer (electrical)
- Electrical power spectrum by the means of spectrum analyzer

Measurable dependent electrical quantities comprise:

- Inductance
- Capacitance
- Electrical impedance defined as vector sum of electrical resistance and electrical reactance
- Electrical admittance, the reciprocal of electrical impedance

- Phase between current and voltage and related power factor
- Electrical spectral density
- Electrical phase noise
- Electrical amplitude noise
- Transconductance
- Transimpedance
- Electrical power gain
- Voltage gain
- Current gain
- Frequency

10

LIGHT

Light is electromagnetic radiation within a certain portion of the electromagnetic spectrum. The word usually refers to visible light, which is visible to the human eye and is responsible for the sense of sight. Visible light is usually defined as having a wavelengthin the range of 400 nanometres (nm), or 400×10 m, to 700 nanometres – between the infrared (with longer wavelengths) and theultraviolet (with shorter wavelengths). Often, infrared and ultraviolet are also called light.

The main source of light on Earth is the Sun. Sunlight provides the energy that green plants use to create sugars mostly in the form of starches, which release energy into the living things that digest them. This process of photosynthesis provides virtually all the energy used by living things. Historically, another important source of light for humans has been fire, from ancient campfires to modern kerosene lamps. With the development of electric lights and of power systems, electric lighting has all but replaced firelight. Some species of animals generate their own light, called bioluminescence. For example, fireflies use light to locate mates, and vampire squids use it to hide themselves from prey.

Primary properties of visible light are intensity, propagation direction, frequency or wavelength spectrum, and polarisation, while itsspeed in a vacuum, 299, 792, 458 meters per second, is one of the fundamental constants of nature. Visible light, as with all types of electromagnetic radiation (EMR), is experimentally found to always move at this speed in vacuum.

In physics, the term light sometimes refers to electromagnetic radiation of any wavelength, whether visible or not. In this sense, gamma rays, X-rays, microwaves and radio waves are also light. Like all types of light, visible light is emitted and absorbed in tiny "packets" called photons, and exhibits properties of both waves and particles. This property is referred to as the wave–particle duality. The study of light, known as optics, is an important research area in modern physics.

ELECTROMAGNETIC SPECTRUM AND VISIBLE LIGHT

Generally, EM radiation, or EMR (the designation 'radiation' excludes static electric and magnetic and near fields) is classified by wavelength into radio, microwave, infrared, the visible region that we perceive as light, ultraviolet, X-rays and gamma rays.

The behaviour of EMR depends on its wavelength. Higher frequencies have shorter wavelengths, and lower frequencies have longer wavelengths. When EMR interacts with single atoms and molecules, its behaviour depends on the amount of energy per quantum it carries.

EMR in the visible light region consists of quanta (called photons) that are at the lower end of the energies that are capable of causing electronic excitation within molecules, which lead to changes in the bonding or chemistry of the molecule. At the lower end of the visible light spectru, EMR becomes invisible to humans (infrared) because its photons no longer have enough individual energy to cause a lasting molecular change (a change in conformation) in the visual molecule retinal in the human retina, which change triggers the sensation of vision.

There exist animals that are sensitive to various types of infrared, but not by means of quantum-absorption. Infrared sensing in snakes depends on a kind of natural thermal imaging, in which tiny packets of cellular water are raised in temperature by the infrared radiation. EMR in this range causes molecular vibration and heating effects, which is how these animals detect it.

Above the range of visible light, ultraviolet light becomes invisible to humans, mostly because it is absorbed by the cornea below 360 nanometers and the internal lens below 400. Furthermore, the rods and cones located in the retina of the human eye cannot detect the very short (below 360 nm) ultraviolet wavelengths, and are in fact damaged by ultraviolet.

Many animals with eyes that do not require lenses (such as insects and shrimp) are able to detect ultraviolet, by quantum photon-absorption mechanisms, in much the same chemical way that humans detect visible light. Various sources define visible light as narrowly as 420 to 680 to as broadly as 380 to 800 nm. Under ideal laboratory conditions, people can see infrared up to at least 1050 nm, children and young adults may perceive ultraviolet wavelengths down to about 310 to 313 nm.

Speed of light

The speed of light in a vacuum is defined to be exactly 299, 792, 458 m/s (approximately 186, 282 miles per second). The fixed value of the speed of light in SI units results from the fact that the metre is now defined in terms of the speed of light. All forms of electromagnetic radiation move at exactly this same speed in vacuum.

Different physicists have attempted to measure the speed of light throughout history. Galileo attempted to measure the speed of light in the seventeenth century. An early experiment to measure the speed of light was conducted by Ole Rømer, a Danish physicist, in 1676. Using a telescope, Rømer observed the motions of Jupiter and one of itsmoons, Io. Noting discrepancies in the apparent period of Io's orbit, he calculated that light takes about 22 minutes to traverse the diameter of Earth's orbit. However, its size was not known at that time. If Rømer had known the diameter of the Earth's orbit, he would have calculated a speed of 227, 000, 000 m/s.

Another, more accurate, measurement of the speed of light was performed in Europe by Hippolyte Fizeau in 1849. Fizeau directed a beam of light at a mirror several kilometers away. A rotating cog wheel was placed in the path of the light beam as it traveled from the source, to the mirror and then returned to its origin. Fizeau found that at a certain rate of rotation, the beam would pass through one gap in the wheel on the way out and the next gap on the way back. Knowing the distance to the mirror, the number of teeth on the wheel, and the rate of rotation, Fizeau was able to calculate the speed of light as 313, 000, 000 m/s.

Léon Foucault used an experiment which used rotating mirrors to obtain a value of 298, 000, 000 m/s in 1862. Albert A. Michelson conducted experiments on the speed of light from 1877 until his death in 1931. He refined Foucault's methods in 1926 using improved rotating mirrors to measure the time it took light to make a round trip from Mount Wilson toMount San Antonio in California. The precise measurements yielded a speed of 299, 796, 000 m/s.

The effective velocity of light in various transparent substances containing ordinary matter, is less than in vacuum. For example, the speed of light in water is about 3/4 of that in vacuum.

Two independent teams of physicists were said to bring light to a "complete standstill" by passing it through a Bose–Einstein condensate of the element rubidium, one team atHarvard University and the Rowland Institute for Science in Cambridge, Mass., and the other at the Harvard–Smithsonian Center for Astrophysics, also in Cambridge. However, the popular description of light being "stopped" in these experiments refers only to light being stored in the excited states of atoms, then re-emitted at an arbitrary later time, as stimulated by a second laser pulse. During the time it had "stopped" it had ceased to be light.

The speed of light in vacuum, commonly denoted c, is a universal physical constant important in many areas of physics. Its value is exactly 299792458 metres per second (H"3. 00×10 m/s), as the length of the metre is defined from this constant and the international standard for time. According to special relativity, c is the maximum speed at which all matter andinformation in the universe can travel. It is the speed at which all massless particles and changes of the associated fields(including electromagnetic radiation such as light and gravitational waves) travel in vacuum. Such particles and waves travel at c regardless of the motion of the source or the inertial frame of reference of the observer. In the theory of relativity, cinterrelates space and time, and also appears in the famous equation of mass–energy equivalence E = mc.

The speed at which light propagates through transparent materials, such as glass or air, is less than c. The ratio between cand the speed v at which light travels in a material is called the refractive index n of the material (n = c / v). For example, forvisible light the refractive index of glass is typically around 1. 5, meaning that light in glass travels at c / 1. 5 H" 200000 km/s; the refractive index of air for visible light is about 1. 0003, so the speed of light in air is about 299700 km/s or 90 km/s slower than c.

For many practical purposes, light and other electromagnetic waves will appear to propagate instantaneously, but for long distances and very sensitive measurements, their finite speed has noticeable effects. In communicating with distant space probes, it can take minutes to hours for a message to get from Earth to the spacecraft, or vice versa. The light seen from stars left them many years ago, allowing the study of the history of the universe by looking at distant objects. The finite speed of light also limits the theoretical maximum speed of computers, since information must be sent within the computer from chip to chip. The speed of light can be used with time of flight measurements to measure large distances to high precision. Ole Rømer first demonstrated in 1676 that light travels at a finite speed (as opposed to instantaneously) by studying the apparent motion of Jupiter's moon Io. In 1865, James Clerk Maxwell proposed that light was an electromagnetic wave, and therefore travelled at the speed c appearing in his theory of electromagnetism. In 1905, Albert Einstein postulated that the speed of light with respect to any inertial frame is independent of the motion of the light source, and explored the consequences of that postulate by deriving the special theory of relativity and showing that the parameter c had relevance outside of the context of light and electromagnetism.

After centuries of increasingly precise measurements, in 1975 the speed of light was known to be 299792458 m/s with ameasurement uncertainty of 4 parts per billion. In 1983, the metre was redefined in the International System of Units (SI) as the distance travelled by light in vacuum in 1/299792458 of a second. As a result, the numerical value of c in metres per second is now fixed exactly by the definition of the metre.

Numerical value, notation, and units

The speed of light in vacuum is usually denoted by a lowercase c, for "constant" or the Latin celeritas (meaning "swiftness, celerity"). Historically, the symbol V was used for as an alternative symbol for the speed of light, introduced by James Clerk Maxwell in 1865. In 1856, Wilhelm Eduard Weber and Rudolf Kohlrausch had used c for a different constant later shown to equal "2 times the speed of light in vacuum. In 1894, Paul Drude redefined c with its modern meaning. Einstein used V in his original German-language papers on special relativity in 1905, but in 1907 he switched to c, which by then had become the standard symbol for the speed of light.

Sometimes c is used for the speed of waves in any material medium, and c0 for the speed of light in vacuum. This subscripted notation, which is endorsed in official SI literature, has the same form as other related constants: namely, ì0 for the vacuum permeability or magnetic constant, å0 for the vacuum permittivity or electric constant, and Z0 for theimpedance of free space. This article uses c exclusively for the speed of light in vacuum.

Since 1983, the metre has been defined in the International System of Units (SI) as the distance light travels in vacuum in D 299792458 of a second. This definition fixes the speed of light in vacuum at exactly 299792458 m/s. As a dimensional physical constant, the numerical value of c is different for different unit systems. In branches of physics in which c appears often, such as in relativity, it is common to use systems of natural units of measurement or the geometrized unit system where c = 1. Using these units, c does not appear explicitly because multiplication or division by 1 does not affect the result.

Fundamental role in physics

The speed at which light waves propagate in vacuum is independent both of the motion of the wave source and of the inertial frame of reference of the observer. This invariance of the speed of light was postulated by Einstein in 1905, after being motivated by Maxwell's theory of electromagnetism and the lack of evidence for the luminiferous aether; it has since been consistently confirmed by many experiments. It is only possible to verify experimentally that the two-way speed of light (for example, from a source to a mirror and back again) is frame-independent, because it is impossible to measure the one-way speed of light (for example, from a source to a distant detector) without some convention as to how clocks at the source and at the detector should be synchronized. However, by adopting Einstein synchronization for the clocks, the one-way speed of light becomes equal to the two-way speed of light by definition. The special theory of relativity explores the consequences of this invariance of c with the assumption that the laws of physics are the same in all inertial frames of reference. One consequence is that c is the speed at which all massless particles and waves, including light, must travel in vacuum.

Special relativity has many counterintuitive and experimentally verified implications. These include the equivalence of mass and energy(E = mc), length contraction (moving objects shorten), and time dilation (moving clocks run more slowly). The factor ã by which lengths contract and times dilate is known as the Lorentz factor and is given by ã = (1 " v/c), where v is the speed of the object. The difference of ãfrom 1 is negligible for speeds much slower than c, such as most everyday speeds—in which case special relativity is closely approximated byGalilean relativity—but it increases at relativistic speeds and diverges to infinity as v approaches c.

The results of special relativity can be summarized by treating space and time as a unified structure known as spacetime (with c relating the units of space and time), and requiring that physical theories satisfy a special symmetry called Lorentz invariance, whose mathematical formulation contains the parameter c. Lorentz invariance is an almost universal assumption for modern physical theories, such as quantum electrodynamics, quantum chromodynamics, the Standard Model of particle physics, and general relativity. As such, the parameter c is ubiquitous in modern physics, appearing in many contexts that are unrelated to light. For example, general relativity predicts that c is also the speed of gravity and of gravitational waves. In non-inertial frames of reference (gravitationally curved space or accelerated reference frames), the local speed of light is constant and equal to c, but the speed of light along a trajectory of finite length can differ from c, depending on how distances and times are defined.

It is generally assumed that fundamental constants such as c have the same value throughout spacetime, meaning that they do not depend on location and do not vary with time. However, it has been suggested in various theories that the speed of light may have changed over time. No conclusive evidence for such changes has been found, but they remain the subject of ongoing research.

It also is generally assumed that the speed of light is isotropic, meaning that it has the same value regardless of the direction in which it is measured. Observations of the emissions from nuclear energy levels as a function of the orientation of the emitting nuclei in a magnetic field, and of rotating optical resonators have put stringent limits on the possible two-way anisotropy.

Upper limit on speeds

According to special relativity, the energy of an object with rest mass m and speed v is given by ãmc, where ã is the Lorentz factor defined above. When v is zero, ã is equal to one, giving rise to the famous E = mc formula for mass–energy equivalence. The ã factor approaches infinity as v approaches c, and it would take an infinite amount of energy to accelerate an object with mass to the speed of light. The speed of light is the upper limit for the speeds of objects with positive rest mass. This is experimentally established in manytests of relativistic energy and momentum. More generally, it is normally impossible for information or energy to travel faster than c. One argument for this follows from the counter-intuitive implication of special relativity known as the relativity of simultaneity. If the spatial distance between two events A and B is greater than the time interval between them multiplied by c then there are frames of reference in which A precedes B, others in which B precedes A, and others in which they are simultaneous. As a result, if something were travelling faster than c relative to an inertial frame of reference, it would be travelling backwards in time relative to another frame, and causality would be violated. In such a frame of reference, an "effect" could be observed before its "cause". Such a violation of causality has never been recorded, and would lead toparadoxes such as the tachyonic antitelephone.

Faster-than-light observations and experiments

There are situations in which it may seem that matter, energy, or information travels at speeds greater than c, but they do not. For example, as is discussed in the propagation of light in a medium section below, many wave velocities can exceed c. For example, the phase velocityof X-rays through most glasses can routinely exceed c, but phase velocity does not determine the velocity at which waves convey information.

If a laser beam is swept quickly across a distant object, the spot of light can move faster than c, although the initial movement of the spot is delayed because of the time it takes light to get to the distant object at the speed c. However, the only physical entities that are moving are the laser and its emitted light, which travels at the speed c from the laser to the various positions of the spot. Similarly, a shadow projected onto a distant object can be made to move faster than c, after a delay in time. In neither case does any matter, energy, or information travel faster than light.

The rate of change in the distance between two objects in a frame of reference with respect to which both are moving (their closing speed) may have a value in excess of c. However, this does not represent the speed of any single object as measured in a single inertial frame.

Certain quantum effects appear to be transmitted instantaneously and therefore faster than c, as in the EPR paradox. An example involves the quantum states of two particles that can be entangled. Until either of the particles is observed, they exist in a superposition of two quantum states. If the particles are separated and one particle's quantum state is observed, the other particle's quantum state is determined instantaneously (i. e., faster than light could travel from one particle to the other). However, it is impossible to control which quantum state the first particle will take on when it is observed, so information cannot be transmitted in this manner.

Another quantum effect that predicts the occurrence of faster-than-light speeds is called the Hartman effect; under certain conditions the time needed for a virtual particle to tunnelthrough a barrier is constant, regardless of the thickness of the barrier. This could result in a virtual particle crossing a large gap faster-than-light. However, no information can be sent using this effect.

So-called superluminal motion is seen in certain astronomical objects, such as the relativistic jets of radio galaxies and quasars. However, these jets are not moving at speeds in excess of the speed of light: the apparent superluminal motion is a projection effect caused by objects moving near the speed of light and approaching Earth at a small angle to the line of sight: since the light which was emitted when the jet was farther away took longer to reach the Earth, the time between two successive observations corresponds to a longer time between the instants at which the light rays were emitted.

In models of the expanding universe, the farther galaxies are from each other, the faster they drift apart. This receding is not due to motion through space, but rather to theexpansion of space itself. For example, galaxies far away from Earth appear to be moving away from the Earth with a speed proportional to their distances. Beyond a boundary called the Hubble sphere, the rate at which their distance from Earth increases becomes greater than the speed of light.

Propagation of light

In classical physics, light is described as a type of electromagnetic wave. The classical behaviour of the electromagnetic field is described by Maxwell's equations, which predict that the speed c with which electromagnetic waves (such as light) propagate through the vacuum is related to the electric constant å0 and the magnetic constant ì0 by the equationc = 1/”å0ì0. In modern quantum physics, the electromagnetic field is described by the theory of quantum electrodynamics (QED). In this theory, light is described by the fundamental excitations (or quanta) of the electromagnetic field, called photons. In QED, photons are massless particles and thus, according to special relativity, they travel at the speed of light in vacuum.

Extensions of QED in which the photon has a mass have been considered. In such a theory, its speed would depend on its frequency, and the invariant speed c of special relativity would then be the upper limit of the speed of light in vacuum. No variation of the speed of light with frequency has been observed in rigorous testing, putting stringent limits on the mass of the photon. The limit obtained depends on the model used: if the massive photon is described by Proca theory, the experimental upper bound for its mass is about 10 grams; if photon mass is generated by a Higgs mechanism, the experimental upper limit is less sharp, m d” 10 eV/c (roughly 2 × 10 g).

Another reason for the speed of light to vary with its frequency would be the failure of special relativity to apply to arbitrarily small scales, as predicted by some proposed theories ofquantum gravity. In 2009, the observation of the spectrum of gamma-ray burst GRB 090510 did not find any difference in the speeds of photons of different energies, confirming that Lorentz invariance is verified at least down to the scale of the Planck length (lP = ”'G/c H” 1. 6163×10 m) divided by 1. 2.

In a medium

In a medium, light usually does not propagate at a speed equal to c; further, different types of light wave will travel at different speeds. The speed at which the individual crests and troughs of a plane wave (a wave filling the whole space, with only one frequency) propagate is called the phase velocity vp. An actual physical signal with a finite extent (a pulse of light) travels at a different speed. The largest part of the pulse travels at the group velocity vg, and its earliest part travels at the front velocity vf.

The phase velocity is important in determining how a light wave travels through a material or from one material to another. It is often represented in terms of a refractive index. The refractive index of a material is defined as the ratio of c to the phase velocity vp in the material: larger indices of refraction indicate lower speeds. The refractive index of a material may depend on the light's frequency, intensity, polarization, or direction of propagation; in many cases, though, it can be treated as a material-dependent constant. The refractive index of air is approximately 1. 0003. Denser media, such as water, glass, and diamond, have refractive indexes of around 1. 3, 1. 5 and 2. 4, respectively, for visible light. In exotic materials like Bose–Einstein condensates near absolute zero, the effective speed of light may be only a few metres per second.

However, this represents absorption and re-radiation delay between atoms, as do all slower-than-c speeds in material substances. As an extreme example of light "slowing" in matter, two independent teams of physicists claimed to bring light to a "complete standstill" by passing it through a Bose–Einstein condensate of the element rubidium, one team at Harvard University and theRowland Institute for Science in Cambridge, Mass., and the other at the Harvard–Smithsonian Center for Astrophysics, also in Cambridge. However, the popular description of light being "stopped" in these experiments refers only to light being stored in the excited states of atoms, then re-emitted at an arbitrarily later time, as stimulated by a second laser pulse. During the time it had "stopped, " it had ceased to be light. This type of behaviour is generally microscopically true of all transparent media which "slow" the speed of light.

In transparent materials, the refractive index generally is greater than 1, meaning that the phase velocity is less than c. In other materials, it is possible for the refractive index to become smaller than 1 for some frequencies; in some exotic materials it is even possible for the index of refraction to become negative. The requirement that causality is not violated implies that the real and imaginary parts of the dielectric constant of any material, corresponding respectively to the index of refraction and to the attenuation coefficient, are linked by the Kramers–Kronig relations. In practical terms, this means that in a material with refractive index less than 1, the absorption of the wave is so quick that no signal can be sent faster than c.

A pulse with different group and phase velocities (which occurs if the phase velocity is not the same for all the frequencies of the pulse) smears out over time, a process known asdispersion. Certain materials have an exceptionally low (or even zero) group velocity for light waves, a phenomenon called slow light, which has been confirmed in various experiments. The opposite, group velocities exceeding c, has also been shown in experiment. It should even be possible for the group velocity to become infinite or negative, with pulses travelling instantaneously or backwards in time.

None of these options, however, allow information to be transmitted faster than c. It is impossible to transmit information with a light pulse any faster than the speed of the earliest part of the pulse (the front velocity). It can be shown that this is (under certain assumptions) always equal to c.

It is possible for a particle to travel through a medium faster than the phase velocity of light in that medium (but still slower than c). When a charged particle does that in a dielectricmaterial, the electromagnetic equivalent of a shock wave, known as Cherenkov radiation, is emitted.

Practical effects of finiteness

The speed of light is of relevance to communications: the one-way and round-trip delay time are greater than zero. This applies from small to astronomical scales. On the other hand, some techniques depend on the finite speed of light, for example in distance measurements.

SMALL SCALES

In supercomputers, the speed of light imposes a limit on how quickly data can be sent between processors. If a processor operates at 1 gigahertz, a signal can only travel a maximum of about 30 centimetres (1 ft) in a single cycle. Processors must therefore be placed close to each other to minimize communication latencies; this can cause difficulty with cooling. If clock frequencies continue to increase, the speed of light will eventually become a limiting factor for the internal design of single chips.

Large distances on Earth

For example, given the equatorial circumference of the Earth is about 40075 km and c about 300000 km/s, the theoretical shortest time for a piece of information to travel half the globe along the surface is about 67 milliseconds. When light is travelling around the globe in an optical fibre, the actual transit time is longer, in part because the speed of light is slower by about 35% in an optical fibre, depending on its refractive index n. Furthermore, straight lines rarely occur in global communications situations, and delays are created when the signal passes through an electronic switch or signal regenerator.

Spaceflights and astronomy

Similarly, communications between the Earth and spacecraft are not instantaneous. There is a brief delay from the source to the receiver, which becomes more noticeable as distances increase. This delay was significant for communications between ground control and Apollo 8 when it became the first manned spacecraft to orbit the Moon: for every question, the ground control station had to wait at least three seconds for the answer to arrive. The communications delay between Earth and Mars can vary between five and twenty minutes depending upon the relative positions of the two planets. As a consequence of this, if a robot on the surface of Mars were to encounter a problem, its human controllers would not be aware of it until at least five minutes later, and possibly up to twenty minutes later; it would then take a further five to twenty minutes for instructions to travel from Earth to Mars.

NASA must wait several hours for information from a probe orbiting Jupiter, and if it needs to correct a navigation error, the fix will not arrive at the spacecraft for an equal amount of time, creating a risk of the correction not arriving in time.

Receiving light and other signals from distant astronomical sources can even take much longer. For example, it has taken 13 billion (13×10) years for light to travel to Earth from the faraway galaxies viewed in the Hubble Ultra Deep Field images. Those photographs, taken today, capture images of the galaxies as they appeared 13 billion years ago, when the universe was less than a billion years old. The fact that more distant objects appear to be younger, due to the finite speed of light, allows astronomers to infer theevolution of stars, of galaxies, and of the universe itself.

Astronomical distances are sometimes expressed in light-years, especially in popular science publications and media. A light-year is the distance light travels in one year, around 9461 billion kilometres, 5879 billion miles, or 0. 3066 parsecs. In round figures, a light year is nearly 10 trillion kilometres or nearly 6 trillion miles. Proxima Centauri, the closest star to Earth after the Sun, is around 4. 2 light-years away.

Distance measurement

Radar systems measure the distance to a target by the time it takes a radio-wave pulse to return to the radar antenna after being reflected by the target: the distance to the target is half the round-trip transit time multiplied by the speed of light. A Global Positioning System (GPS) receiver measures its distance to GPS satellites based on how long it takes for a radio signal to arrive from each satellite, and from these distances calculates the receiver's position. Because light travels about 300000 kilometres (186000 mi) in one second, these measurements of small fractions of a second must be very precise. The Lunar Laser Ranging Experiment, radar astronomy and the Deep Space Network determine distances to the Moon, planets and spacecraft, respectively, by measuring round-trip transit times.

High-frequency trading

The speed of light has become important in high-frequency trading, where traders seek to gain minute advantages by delivering their trades to exchanges fractions of a second ahead of other traders. For example traders have been switching to microwave communications between trading hubs, because of the advantage which microwaves travelling at near to the speed of light in air, have over fibre optic signals which travel 30–40% slower at the speed of light through glass.

Measurement

There are different ways to determine the value of c. One way is to measure the actual speed at which light waves propagate, which can be done in various astronomical and earth-based setups. However, it is also possible to determine c from other physical laws where it appears, for example, by determining the values of the electromagnetic constantså0 and ì0 and using their relation to c. Historically, the most accurate results have been obtained by separately determining the frequency and wavelength of a light beam, with their product equalling c.

In 1983 the metre was defined as "the length of the path travelled by light in vacuum during a time interval of D 299792458 of a second", fixing the value of the speed of light at299792458 m/s by definition, as described below. Consequently, accurate measurements of the speed of light yield an accurate realization of the metre rather than an accurate value of c.

Astronomical measurements

Outer space is a convenient setting for measuring the speed of light because of its large scale and nearly perfect vacuum. Typically, one measures the time needed for light to traverse some reference distance in the solar system, such as the radiusof the Earth's orbit. Historically, such measurements could be made fairly accurately, compared to how accurately the length of the reference distance is known in Earth-based units. It is customary to express the results in astronomical units (AU) per day.

Ole Christensen Rømer used an astronomical measurement to make the first quantitative estimate of the speed of light. When measured from Earth, the periods of moons orbiting a distant planet are shorter when the Earth is approaching the planet than when the Earth is receding from it. The distance travelled by light from the planet (or its moon) to Earth is shorter when the Earth is at the point in its orbit that is closest to its planet than when the Earth is at the farthest point in its orbit, the difference in distance being the diameter of the Earth's orbit around the Sun. The observed change in the moon's orbital period is caused by the difference in the time it takes light to traverse the shorter or longer distance. Rømer observed this effect for Jupiter's innermost moon Io and deduced that light takes 22 minutes to cross the diameter of the Earth's orbit.

Another method is to use the aberration of light, discovered and explained by James Bradley in the 18th century. This effect results from the vector addition of the velocity of light arriving from a distant source (such as a star) and the velocity of its observer. A moving observer thus sees the light coming from a slightly different direction and consequently sees the source at a position shifted from its original position. Since the direction of the Earth's velocity changes continuously as the Earth orbits the Sun, this effect causes the apparent position of stars to move around. From the angular difference in the position of stars (maximally 20. 5 arcseconds) it is possible to express the speed of light in terms of the Earth's velocity around the Sun, which with the known length of a year can be converted to the time needed to travel from the Sun to the Earth. In 1729, Bradley used this method to derive that light travelled 10, 210 times faster than the Earth in its orbit (the modern figure is 10, 066 times faster) or, equivalently, that it would take light 8 minutes 12 seconds to travel from the Sun to the Earth.

Astronomical unit

An astronomical unit (AU) is approximately the average distance between the Earth and Sun. It was redefined in 2012 as exactly 149597870700 m. Previously the AU was not based on the International System of Units but in terms of the gravitational force exerted by the Sun in the framework of classical mechanics. The current definition uses the recommended value in metres for the previous definition of the astronomical unit, which was determined by measurement. This redefinition is analogous to that of the metre, and likewise has the effect of fixing the speed of light to an exact value in astronomical units per second (via the exact speed of light in metres per second).

Previously, the inverse of c expressed in seconds per astronomical unit was measured by comparing the time for radio signals to reach different spacecraft in the Solar System, with their position calculated from the gravitational effects of the Sun and various planets. By combining many such measurements, a best fit value for the light time per unit distance could be obtained. For example, in 2009, the best estimate, as approved by the International Astronomical Union (IAU), was:

light time for unit distance: 499. 004783836(10) s

c = 0. 00200398880410(4) AU/s = 173. 144632674(3) AU/day.

The relative uncertainty in these measurements is 0. 02 parts per billion (2×10), equivalent to the uncertainty in Earth-based measurements of length by interferometry. Since the metre is defined to be the length travelled by light in a certain time interval, the measurement of the light time in terms of the previous definition of the astronomical unit can also be interpreted as measuring the length of an AU (old definition) in metres.

Time of flight techniques

A method of measuring the speed of light is to measure the time needed for light to travel to a mirror at a known distance and back. This is the working principle behind the Fizeau–Foucault apparatus developed by Hippolyte Fizeau and Léon Foucault.

The setup as used by Fizeau consists of a beam of light directed at a mirror 8 kilometres (5 mi) away. On the way from the source to the mirror, the beam passes through a rotating cogwheel. At a certain rate of rotation, the beam passes through one gap on the way out and another on the way back, but at slightly higher or lower rates, the beam strikes a tooth and does not pass through the wheel. Knowing the distance between the wheel and the mirror, the number of teeth on the wheel, and the rate of rotation, the speed of light can be calculated.

The method of Foucault replaces the cogwheel by a rotating mirror. Because the mirror keeps rotating while the light travels to the distant mirror and back, the light is reflected from the rotating mirror at a different angle on its way out than it is on its way back. From this difference in angle, the known speed of rotation and the distance to the distant mirror the speed of light may be calculated.

Nowadays, using oscilloscopes with time resolutions of less than one nanosecond, the speed of light can be directly measured by timing the delay of a light pulse from a laser or an LED reflected from a mirror. This method is less precise (with errors of the order of 1%) than other modern techniques, but it is sometimes used as a laboratory experiment in college physics classes.

Electromagnetic constants

An option for deriving c that does not directly depend on a measurement of the propagation of electromagnetic waves is to use the relation between c and the vacuum permittivityå0 and vacuum permeability ì0 established by Maxwell's theory: c = 1/(å0ì0). The vacuum permittivity may be determined by measuring the capacitance and dimensions of acapacitor, whereas the value of the vacuum permeability is fixed at exactly 4ð×10 H·m through the definition of the ampere. Rosa and Dorsey used this method in 1907 to find a value of 299710±22 km/s.

Cavity resonance

Another way to measure the speed of light is to independently measure the frequency f and wavelength ë of an electromagnetic wave in vacuum. The value of c can then be found by using the relation c = fë. One option is to measure the resonance frequency of a cavity resonator. If the dimensions of the resonance cavity are also known, these can be used determine the wavelength of the wave. In 1946, Louis Essen and A. C. Gordon-Smith established the frequency for a variety of normal modes of microwaves of a microwave cavity of precisely known dimensions. The dimensions were established to an accuracy of about ±0. 8 ìm using gauges calibrated by interferometry. As the wavelength of the modes was known from the geometry of the cavity and from electromagnetic theory, knowledge of the associated frequencies enabled a calculation of the speed of light.

The Essen–Gordon-Smith result, 299792±9 km/s, was substantially more precise than those found by optical techniques. By 1950, repeated measurements by Essen established a result of 299792. 5±3. 0 km/s.

A household demonstration of this technique is possible, using a microwave oven and food such as marshmallows or margarine: if the turntable is removed so that the food does not move, it will cook the fastest at the antinodes (the points at which the wave amplitude is the greatest), where it will begin to melt. The distance between two such spots is half the wavelength of the microwaves; by measuring this distance and multiplying the wavelength by the microwave frequency (usually displayed on the back of the oven, typically 2450 MHz), the value of c can be calculated, "often with less than 5% error".

Interferometry

Interferometry is another method to find the wavelength of electromagnetic radiation for determining the speed of light. Acoherent beam of light (e. g. from a laser), with a known frequency (f), is split to follow two paths and then recombined. By adjusting the path length while observing the interference pattern and carefully measuring the change in path length, the wavelength of the light (ë) can be determined. The speed of light is then calculated using the equation c = ëf.

Before the advent of laser technology, coherent radio sources were used for interferometry measurements of the speed of light. However interferometric determination of wavelength becomes less precise with wavelength and the experiments were thus limited in precision by the long wavelength (~0. 4 cm) of the radiowaves. The precision can be improved by using light with a shorter wavelength, but then it becomes difficult to directly measure the frequency of the light. One way around this problem is to start with a low frequency signal of which the frequency can be precisely measured, and from this signal progressively synthesize higher frequency signals whose frequency can then be linked to the original signal. A laser can then be locked to the frequency, and its wavelength can be determined using interferometry. This technique was due to a group at the National Bureau of Standards (NBS) (which later became NIST). They used it in 1972 to measure the speed of light in vacuum with a fractional uncertainty of3. 5×10.

History

Until the early modern period, it was not known whether light travelled instantaneously or at a very fast finite speed. The first extant recorded examination of this subject was in ancient Greece. The ancient Greeks, Muslim scholars and classical European scientists long debated this until Rømer provided the first calculation of the speed of light. Einstein's Theory of Special Relativity concluded that the speed of light is constant regardless of one's frame of reference. Since then, scientists have provided increasingly accurate measurements.

Early history

Empedocles (c. 490–430 BC) was the first to claim that light has a finite speed. He maintained that light was something in motion, and therefore must take some time to travel. Aristotle argued, to the contrary, that "light is due to the presence of something, but it is not a movement". Euclid andPtolemy advanced Empedocles' emission theory of vision, where light is emitted from the eye, thus enabling sight. Based on that theory, Heron of Alexandria argued that the speed of light must be infinite because distant objects such as stars appear immediately upon opening the eyes.

Early Islamic philosophers initially agreed with the Aristotelian view that light had no speed of travel. In 1021, Alhazen (Ibn al-Haytham) published the Book of Optics, in which he presented a series of arguments dismissing the emission theory of vision in favour of the now accepted intromission theory, in which light moves from an object into the eye. This led Alhazen to propose that light must have a finite speed, and that the speed of light is variable, decreasing in denser bodies. He argued that light is substantial matter, the propagation of which requires time, even if this is hidden from our senses. Also in the 11th century, Abû Rayhân al-Bîrûnî agreed that light has a finite speed, and observed that the speed of light is much faster than the speed of sound.

In the 13th century, Roger Bacon argued that the speed of light in air was not infinite, using philosophical arguments backed by the writing of Alhazen and Aristotle. In the 1270s, Witelo considered the possibility of light travelling at infinite speed in vacuum, but slowing down in denser bodies.

In the early 17th century, Johannes Kepler believed that the speed of light was infinite, since empty space presents no obstacle to it. René Descartes argued that if the speed of light were to be finite, the Sun, Earth, and Moon would be noticeably out of alignment during a lunar eclipse. Since such misalignment had not been observed, Descartes concluded the speed of light was infinite. Descartes speculated that if the speed of light were found to be finite, his whole system of philosophy might be demolished. In Descartes' derivation of Snell's law, he assumed that even though the speed of light was instantaneous, the more dense the medium, the faster was light's speed. Pierre de Fermat derived Snell's law using the opposing assumption, the more dense the medium the slower light traveled. Fermat also argued in support of a finite speed of light.

First measurement attempts

In 1629, Isaac Beeckman proposed an experiment in which a person observes the flash of a cannon reflecting off a mirror about one mile (1. 6 km) away. In 1638, Galileo Galileiproposed an experiment, with an apparent claim to having performed it some years earlier, to measure the speed of light by observing the delay between uncovering a lantern and its perception some distance away. He was unable to distinguish whether light travel was instantaneous or not, but concluded that if it were not, it must nevertheless be extraordinarily rapid. Galileo's experiment was carried out by the Accademia del Cimento of Florence, Italy, in 1667, with the lanterns separated by about one mile, but no delay was observed. The actual delay in this experiment would have been about 11 microseconds.

The first quantitative estimate of the speed of light was made in 1676 by Rømer. From the observation that the periods of Jupiter's innermost moon Io appeared to be shorter when the Earth was approaching Jupiter than when receding from it, he concluded that light travels at a finite speed, and estimated that it takes light 22 minutes to cross the diameter of Earth's orbit. Christiaan Huygens combined this estimate with an estimate for the diameter of the Earth's orbit to obtain an estimate of speed of light of220000 km/s, 26% lower than the actual value.

In his 1704 book Opticks, Isaac Newton reported Rømer's calculations of the finite speed of light and gave a value of "seven or eight minutes" for the time taken for light to travel from the Sun to the Earth (the modern value is 8 minutes 19 seconds). Newton queried whether Rømer's eclipse shadows were coloured; hearing that they were not, he concluded the different colours travelled at the same speed. In 1729, James Bradley discovered stellar aberration. From this effect he determined that light must travel 10, 210 times faster than the Earth in its orbit (the modern figure is 10, 066 times faster) or, equivalently, that it would take light 8 minutes 12 seconds to travel from the Sun to the Earth.

Connections with electromagnetism

In the 19th century Hippolyte Fizeau developed a method to determine the speed of light based on time-of-flight measurements on Earth and reported a value of 315000 km/s. His method was improved upon by Léon Foucault who obtained a value of 298000 km/s in 1862. In the year 1856, Wilhelm Eduard Weber and Rudolf Kohlrausch measured the ratio of the electromagnetic and electrostatic units of charge, 1/"å0ì0, by discharging a Leyden jar, and found that its numerical value was very close to the speed of light as measured directly by Fizeau. The following year Gustav Kirchhoff calculated that an electric signal in a resistanceless wire travels along the wire at this speed. In the early 1860s, Maxwell showed that, according to the theory of electromagnetism he was working on, electromagnetic waves propagate in empty space at a speed equal to the above Weber/Kohrausch ratio, and drawing attention to the numerical proximity of this value to the speed of light as measured by Fizeau, he proposed that light is in fact an electromagnetic wave.

"Luminiferous aether"

It was thought at the time that empty space was filled with a background medium called the luminiferous aether in which the electromagnetic field existed. Some physicists thought that this aether acted as a preferred frame of reference for the propagation of light and therefore it should be possible to measure the motion of the Earth with respect to this medium, by measuring the isotropy of the speed of light. Beginning in the 1880s several experiments were performed to try to detect this motion, the most famous of which is the experiment performed by Albert A. Michelson andEdward W. Morley in 1887. The detected motion was always less than the observational error. Modern experiments indicate that the two-way speed of light is isotropic (the same in every direction) to within 6 nanometres per second. Because of this experiment Hendrik Lorentzproposed that the motion of the apparatus through the aether may cause the apparatus to contract along its length in the direction of motion, and he further assumed, that the time variable for moving systems must also be changed accordingly ("local time"), which led to the formulation of theLorentz transformation. Based on Lorentz's aether theory, Henri Poincaré (1900) showed that this local time (to first order in v/c) is indicated by clocks moving in the aether, which are synchronized under the assumption of constant light speed. In 1904, he speculated that the speed of light could be a limiting velocity in dynamics, provided that the assumptions of Lorentz's theory are all confirmed. In 1905, Poincaré brought Lorentz's aether theory into full observational agreement with the principle of relativity.

Special relativity

In 1905 Einstein postulated from the outset that the speed of light in vacuum, measured by a non-accelerating observer, is independent of the motion of the source or observer. Using this and the principle of relativity as a basis he derived the special theory of relativity, in which the speed of light in vacuum c featured as a fundamental constant, also appearing in contexts unrelated to light. This made the concept of the stationary aether (to which Lorentz and Poincaré still adhered) useless and revolutionized the concepts of space and time.

Increased accuracy of c and redefinition of the metre and second

In the second half of the 20th century much progress was made in increasing the accuracy of measurements of the speed of light, first by cavity resonance techniques and later by laser interferometer techniques. These were aided by new, more precise, definitions of the metre and second. In 1950, Louis Essen determined the speed as 299792. 5±1 km/s, using cavity resonance. This value was adopted by the 12th General Assembly of the Radio-Scientific Union in 1957. In 1960, the metre was redefined in terms of the wavelength of a particular spectral line of krypton-86, and, in 1967, the second was redefined in terms of the hyperfine transition frequency of the ground state of caesium-133.

In 1972, using the laser interferometer method and the new definitions, a group at NBS in Boulder, Colorado determined the speed of light in vacuum to bec = 299792456. 2±1. 1 m/s. This was 100 times less uncertain than the previously accepted value. The remaining uncertainty was mainly related to the definition of the metre. As similar experiments found comparable results for c, the 15th Conférence Générale des Poids et Mesures (CGPM) in 1975 recommended using the value299792458 m/s for the speed of light.

Defining the speed of light as an explicit constant

In 1983 the 17th CGPM found that wavelengths from frequency measurements and a given value for the speed of light are more reproducible than the previous standard. They kept the 1967 definition of second, so the caesium hyperfine frequency would now determine both the second and the metre. To do this, they redefined the metre as: "The metre is the length of the path travelled by light in vacuum during a time interval of 1/299792458 of a second. " As a result of this definition, the value of the speed of light in vacuum is exactly 299792458 m/s and has become a defined constant in the SI system of units. Improved experimental techniques that prior to 1983 would have measured the speed of light, no longer affect the known value of the speed of light in SI units, but instead allow a more precise realization of the metre by more accurately measuring the wavelength of Krypton-86 and other light sources.

In 2011, the CGPM stated its intention to redefine all seven SI base units using what it calls "the explicit-constant formulation", where each "unit is defined indirectly by specifying explicitly an exact value for a well-recognized fundamental constant", as was done for the speed of light. It proposed a new, but completely equivalent, wording of the metre's definition: "The metre, symbol m, is the unit of length; its magnitude is set by fixing the numerical value of the speed of light in vacuum to be equal to exactly 299792458 when it is expressed in the SI unit m s. " This is one of the proposed changes to be incorporated in the next revision of the SI also termed the New SI.

Optics

The study of light and the interaction of light and matter is termed optics. The observation and study of optical phenomena such as rainbows and the aurora borealis offer many clues as to the nature of light.

Optics is the branch of physics which involves the behaviour and properties of light, including its interactions with matter and the construction of instruments that use or detect it. Optics usually describes the behaviour of visible, ultraviolet, and infrared light. Because light is an electromagnetic wave, other forms of electromagnetic radiation such as X-rays, microwaves, and radio waves exhibit similar properties.

Most optical phenomena can be accounted for using the classical electromagnetic description of light. Complete electromagnetic descriptions of light are, however, often difficult to apply in practice. Practical optics is usually done using simplified models. The most common of these, geometric optics, treats light as a collection of rays that travel in straight lines and bend when they pass through or reflect from surfaces. Physical optics is a more comprehensive model of light, which includes wave effects such as diffraction and interference that cannot be accounted for in geometric optics. Historically, the ray-based model of light was developed first, followed by the wave model of light. Progress in electromagnetic theory in the 19th century led to the discovery that light waves were in fact electromagnetic radiation.

Some phenomena depend on the fact that light has both wave-like and particle-like properties. Explanation of these effects requiresquantum mechanics. When considering light's particle-like properties, the light is modelled as a collection of particles called "photons". Quantum optics deals with the application of quantum mechanics to optical systems.

Optical science is relevant to and studied in many related disciplines including astronomy, various engineering fields, photography, andmedicine (particularly ophthalmology and optometry). Practical applications of optics are found in a variety of technologies and everyday objects, including mirrors, lenses, optical telescopes, microscopes, lasers, and fibre optics.

Refraction

Refraction is the bending of light rays when passing through a surface between one transparent material and another. It is described by Snell's Law: $n_1 \sin \theta_1 = n_2 \sin \theta_2$.

where θ_1 is the angle between the ray and the surface normal in the first medium θ_2, is the angle between the ray and the surface normal in the second medium, and n1 and n2 are the indices of refraction, $n = 1$ in a vacuum and $n > 1$ in a transparent substance.

When a beam of light crosses the boundary between a vacuum and another medium, or between two different media, the wavelength of the light changes, but the frequency remains constant. If the beam of light is not orthogonal (or rather normal) to the boundary, the change in wavelength results in a change in the direction of the beam. This change of direction is known as refraction.

The refractive quality of lenses is frequently used to manipulate light in order to change the apparent size of images. Magnifying glasses, spectacles, contact lenses, microscopes and refracting telescopes are all examples of this manipulation.

Refraction is the change in direction of propagation of a wave due to a change in its transmission medium.

The phenomenon is explained by the conservation of energy and conservation of momentum. Due to change of medium, the phase velocity of the wave is changed but its frequency remains constant. This is most commonly observed when a wave passes from onemedium to another at any angle other than 0° from the normal. Refraction of light is the most commonly observed phenomenon, but any type of wave can refract when it interacts with a medium, for example when sound waves pass from one medium into another or when water waves move into water of a different depth. Refraction is described by Snell's law, which states that for a given pair of media and a wave with a single frequency, the ratio of the sines of the angle of incidence è1 and angle of refraction è2 is equivalent to the ratio of phase velocities (v_1 / v_2) in the two media, or equivalently, to the opposite ratio of the indices of refraction (n_2 / n_1):

$$\frac{\sin\theta_1}{\sin\theta_2} = \frac{v_1}{v_2} = \frac{n_2}{n_1}.$$

In general, the incident wave is partially refracted and partially reflected; the details of this behavior are described by the Fresnel equations.

Explanation

In optics, refraction is a phenomenon that often occurs when waves travel from a medium with a given refractive index to a medium with another at an oblique angle. At the boundary between the media, the wave's phase velocity is altered, usually causing a change in direction. Its wavelength increases or decreases but its frequency remains constant. For example, a light ray will refract as it enters and leaves glass, assuming there is a change in refractive index. A ray traveling along the normal (perpendicular to the boundary) will change speed, but not direction. Refraction still occurs in this case. Understanding of this concept led to the invention of lenses and therefracting telescope.

Refraction can be seen when looking into a bowl of water. Air has a refractive index of about 1. 0003, and water has a refractive index of about 1. 3330. If a person looks at a straight object, such as a pencil or straw, which is placed at a slant, partially in the water, the object appears to bend at the water's surface. This is due to the bending of light rays as they move from the water to the air. Once the rays reach the eye, the eye traces them back as straight lines (lines of sight). The lines of sight (shown as dashed lines) intersect at a higher position than where the actual rays originated. This causes the pencil to appear higher and the water to appear shallower than it really is. The depth that the water appears to be when viewed from above is known as the apparent depth. This is an important consideration for spearfishingfrom the surface because it will make the target fish appear to be in a different place, and the fisher must aim lower to catch the fish. Conversely, an object above the water has a higher apparent height when viewed from below the water. The opposite correction must be made by an archer fish. For small angles of incidence (measured from the normal, when sin è is approximately the same as tan è), the ratio of apparent to real depth is the ratio of the refractive indexes of air to that of water. But as the angle of incidence approaches 90, the apparent depth approaches zero, albeit reflection increases, which limits observation at high angles of incidence. Conversely, the apparent height approaches infinity as the angle of incidence (from below) increases, but even earlier, as the angle of total internal reflection is approached, albeit the image also fades from view as this limit is approached.

The diagram on the right shows an example of refraction in water waves. Ripples travel from the left and pass over a shallower region inclined at an angle to the wavefront. The waves travel slower in the more shallow water, so the wavelength decreases and the wave bends at the boundary. The dotted line represents the normal to the boundary. The dashed line represents the original direction of the waves. This phenomenon explains why waves on a shoreline tend to strike the shore close to a perpendicular angle. As the waves travel from deep water into shallower water near the shore, they are refracted from their original direction of travel to an angle more normal to the shoreline. Refraction is also responsible for rainbows and for the splitting of white light into a rainbow-spectrum as it passes through a glass prism. Glass has a higher refractive index than air. When a beam of white light passes from air into a material having an index of refraction that varies with frequency, a phenomenon known as dispersion occurs, in which different coloured components of the white light are refracted at different angles, i. e., they bend by different amounts at the interface, so that they become separated. The different colors correspond to different frequencies.

While refraction allows for phenomena such as rainbows, it may also produce peculiar optical phenomena, such as mirages and Fata Morgana. These are caused by the change of the refractive index of air with temperature.

The refractive index of materials can also be nonlinear, as occurs with the Kerr effect when high intensity light leads to a refractive index proportional to the intensity of the incident light.

Recently some metamaterials have been created which have a negative refractive index. With metamaterials, we can also obtain total refraction phenomena when the wave impedances of the two media are matched. There is then no reflected wave.

Also, since refraction can make objects appear closer than they are, it is responsible for allowing water to magnify objects. First, as light is entering a drop of water, it slows down. If the water's surface is not flat, then the light will be bent into a new path. This round shape will bend the light outwards and as it spreads out, the image you see gets larger.

An analogy that is often put forward to explain the refraction of light is as follows: "Imagine a marching band as they march at an oblique angle from pavement (a fast medium) into mud (a slower medium). The marchers on the side that runs into the mud first will slow down first. This causes the whole band to pivot slightly toward the normal (make a smaller angle from the normal). "

Light sources

There are many sources of light. The most common light sources are thermal: a body at a given temperature emits a characteristic spectrum of black-body radiation. A simple thermal source is sunlight, the radiation emitted by the chromosphere of the Sun at around 6, 000 Kelvin peaks in the visible region of the electromagnetic spectrum when plotted in wavelength units and roughly 44% of sunlight energy that reaches the ground is visible. Another example is incandescent light bulbs, which emit only around 10% of their energy as visible light and the remainder as infrared. A common thermal light source in history is the glowing solid particles in flames, but these also emit most of their radiation in the infrared, and only a fraction in the visible spectrum. The peak of the blackbody spectrum is in the deep infrared, at about 10 micrometer wavelength, for relatively cool objects like human beings. As the temperature increases, the peak shifts to shorter wavelengths, producing first a red glow, then a white one, and finally a blue-white colour as the peak moves out of the visible part of the spectrum and into the ultraviolet. These colours can be seen when metal is heated to "red hot" or "white hot". Blue-white thermal emission is not often seen, except in stars (the commonly seen pure-blue colour in a gas flame or a welder's torch is in fact due to molecular emission, notably by CH radicals (emitting a wavelength band around 425 nm, and is not seen in stars or pure thermal radiation).

Atoms emit and absorb light at characteristic energies. This produces "emission lines" in the spectrum of each atom. Emission can bespontaneous, as in light-emitting diodes, gas discharge lamps (such as neon lamps and neon signs, mercury-vapor lamps, etc.), and flames (light from the hot gas itself—so, for example, sodium in a gas flame emits characteristic yellow light). Emission can also be stimulated, as in a laser or a microwave maser.

BIBLIOGRAPHY

- Adkins, C. J. (1968/1983). Equilibrium Thermodynamics, (1st edition 1968), third edition 1983, Cambridge University Press, Cambridge UK, ISBN 0-521-25445-0.
- Ampère, André-Marie. Essai sur la Pilosophie des Sciences. Chez Bachelier.
- Bacon, F. (1620). Novum Organum Scientiarum, translated by Devey, J., P. F. Collier & Son, New York, 1902.
- Baierlein, R. (1999). Thermal Physics. Cambridge University Press. ISBN 978-0-521-65838-6.
- Bailyn, M. (1994). A Survey of Thermodynamics, American Institute of Physics Press, New York, ISBN 0-88318-797-3.
- Born, M. (1949). Natural Philosophy of Cause and Chance, Oxford University Press, London.
- Bryan, G. H. (1907). Thermodynamics. An Introductory Treatise dealing mainly with First Principles and their Direct Applications, B. G. Teubner, Leipzig.
- Chandrasekhar, S. (1961). Hydrodynamic and Hydromagnetic Stability, Oxford University Press, Oxford UK.
- Clark, J. O. E. (2004). The Essential Dictionary of Science. Barnes & Noble Books. ISBN 0-7607-4616-8.
- Draper, John William (1861). A textbook on chemistry. New York: Harper and Sons. p. 178.
- J. Clerk Maxwell (1904). Theory of Heat. Mineola: Dover Publications. pp. 319–20. ISBN 0-486-41735-2.

- J. M. McCarthy and G. S. Soh, 2010, Geometric Design of Linkages, Springer, New York.
- Jack Erjavec "Automotive technology: a systems approach" Delmar Learning 2000 p. 309 ISBN 1-4018-4831-1
- Joseph Stiles Beggs (1983). Kinematics. Taylor & Francis. p. 1. ISBN 0-89116-355-7.
- Kenneth Wark (1977). Thermodynamics (3 ed.). McGraw-Hill. p. 12. ISBN 0-07-068280-1.
- McPherson, pp. 52–55
- McPherson, pp. 55–60
- Reuleaux, F. ; Kennedy, Alex B. W. (1876), The Kinematics of Machinery: Outlines of a Theory of Machines, London: Macmillan
- Russell C. Hibbeler (2009). "Kinematics and kinetics of a particle". Engineering Mechanics: Dynamics (12th ed.). Prentice Hall. p. 298. ISBN 0-13-607791-9. ;
- The authors make the connection between molecular forces of metals and their corresponding physical properties. By extension, this concept would apply to gases as well, though not universally. Cornell (1907) pp. 164–5.
- The work by T. Zelevinski provides another link to latest research about Strontium in this new field of study. See Tanya Zelevinsky (2009). "84Sr—just right for forming a Bose-Einstein condensate". Physics 2: 94. Bibcode:2009PhyOJ... 2... 94Z. doi:10. 1103/physics. 2. 94.
- Theo Mang, Wilfried Dressel "Lubricants and lubrication", Wiley-VCH 2007 ISBN 3-527-31497-0
- Theodore Gray, The Elements: A Visual Exploration of Every Known Atom in the Universe New York: Workman Publishing, 2009 p. 127 ISBN 1-57912-814-9
- Thomas Wallace Wright (1896). Elements of Mechanics Including Kinematics, Kinetics and Statics. E and FN Spon. Chapter 1.

INDEX

S

T

V

W

Z